INTRODUCTION TO CALCULUS

**ADIWES INTERNATIONAL SERIES
IN MATHEMATICS**

A. J. Lohwater, Consulting Editor

INTRODUCTION TO CALCULUS

by

KAZIMIERZ KURATOWSKI

Professor of Mathematics, The University of Warsaw
Member of the Polish Academy of Sciences

TRANSLATED FROM THE POLISH

by

DOCTOR JULIAN MUSIELAK

of the University of Poznań

SECOND EDITION

PERGAMON PRESS

OXFORD · LONDON · EDINBURGH · NEW YORK
TORONTO · SYDNEY · PARIS · BRAUNSCHWEIG

ADDISON-WESLEY PUBLISHING COMPANY

READING, MASSACHUSETTS · MENLO PARK, CALIFORNIA
LONDON · DON MILLS, ONTARIO

Copyright © 1969

PAŃSTWOWE WYDAWNICTWO NAUKOWE, WARSZAWA
(PWN — Polish Scientific Publishers)
First English Edition 1961
Second English Edition 1969

U.S.A. Edition distributed by
Addison-Wesley Publishing Company, Inc.
Reading, Massachusetts, U.S.A.

PERGAMON PRESS
International Series of Monographs in
PURE AND APPLIED MATHEMATICS
Volume 17

Library of Congress Catalog Card Number 68-23037

Printed in Poland

```
QA            Kuratowski,
303             Kazimierz
.K883
1969b         Introduction to
                calculus.
515 K965i2
```

CONTENTS

Preface to the English edition 9
Preface to the Polish edition 10

I. SEQUENCES AND SERIES

§ 1. Introduction

1.1. Various kinds of numbers 11
1.2. The principle of mathematical induction 12
1.3. The Newton binomial formula 15
1.4*. Schwarz inequality 17
1.5. The principle of continuity (Dedekind) 18
1.6. The absolute value of a number 19
1.7. Bounded sets. The upper bound and the lower bound of a set 19
1.8*. The axiomatic treatment of real numbers 21
1.9*. Real numbers as sets of rational numbers 23
Exercises on § 1 25

§ 2. Infinite sequences

2.1. Definition and examples 26
2.2. The notion of limit 28
2.3. Bounded sequences 32
2.4. Operations on sequences 32
2.5. Further properties of the limit 36
2.6. Subsequences 38
2.7. Cauchy theorem 42
2.8. Divergence to ∞ 44
2.9. Examples 46
2.10. The number e 47
2.11*. The sequences of the arithmetic means and of the geometric means of a given sequence 49
Exercises on § 2 52

§ 3. Infinite series

3.1. Definitions and examples 55
3.2. General properties of series 56
3.3. Alternating series. Abel theorem 59
3.4. Series with positive terms. D'Alembert and Cauchy convergence criterions 61

3.5. Applications and examples	64
3.6. Other convergence criteria	66
3.7. Absolutely convergent series	68
3.8. Multiplication of series	71
3.9*. Infinite products	74
Exercises on § 3	79

II. FUNCTIONS

§ 4. Functions and their limits

4.1. Definitions	81
4.2. Monotone functions	83
4.3. One-to-one functions. Inverse functions	85
4.4. Elementary functions	86
4.5. The limit of a function f at a point a	89
4.6. Operations on the limit	93
4.7. Conditions for the existence of the limit	97
Exercises on § 4	101

§ 5. Continuous functions

5.1. Definition	102
5.2. Cauchy characterization of continuity. Geometrical interpretation	104
5.3. Continuity of elementary functions	105
5.4. General properties of continuous functions	109
5.5. Continuity of inverse functions	114
Exercises on § 5	116

§ 6. Sequences and series of functions

6.1. Uniform convergence	118
6.2. Uniformly convergent series	121
6.3. Power series	123
6.4. Approximation of continuous functions by polygonal functions	127
6.5*. The symbolism of mathematical logic	129
Exercises on § 6	137

III. DIFFERENTIAL CALCULUS

§ 7. Derivatives of the first order

7.1. Definitions	138
7.2. Differentiation of elementary functions	142
7.3. Differentiation of inverse functions	147

7.4.	Extrema of functions. Rolle theorem	149
7.5.	Lagrange and Cauchy theorems	152
7.6.	Differentiation of composite functions	156
7.7.	Geometrical interpretation of the sign of a derivative	161
7.8.	Indeterminate expressions	163
7.9.	The derivative of a limit	168
7.10.	The derivative of a power series	169
7.11.	The expansion of the functions $\log(1+x)$ and $\arctan x$ in power series	172
7.12*.	Asymptotes	174
7.13*.	The concept of a differential	175
	Exercises on § 7	178

§ 8. Derivatives of higher orders

8.1.	Definition and examples	180
8.2*.	Differentials of higher order	182
8.3.	Arithmetical operations	184
8.4.	Taylor formula	185
8.5.	Expansions in power series	190
8.6.	A criterion for extrema	194
8.7.	Geometrical interpretation of the second derivative. Points of inflexion	196
	Exercises on § 8	199

IV. INTEGRAL CALCULUS

§ 9. Indefinite integrals

9.1.	Definition	201
9.2.	The integral of the limit. Integrability of continuous functions	204
9.3.	General formulae for integration	205
9.4.	Integration of rational functions	211
9.5.	Integration of irrational functions of the second degree	215
9.6.	Integration of trigonometric functions	219
	Exercises on § 9	223

§ 10. Definite integrals

10.1.	Definition and examples	224
10.2.	Calculation formulae	227
10.3.	Definite integral as a limit of sums	234
10.4.	The integral as an area	237
10.5.	The length of an arc	241
10.6.	The volume and surface area of a solid of revolution	248
10.7.	Two mean-value theorems	253

CONTENTS

- 10.8. Methods of approximate integrations. Lagrange interpolation — 256
- 10.9. Wallis formula — 259
- 10.10. Stirling formula — 261
- 10.11*. Riemann integral. Upper and lower Darboux integrals — 262
- Exercises on § 10 — 270

§ 11. Improper integrals and their connection with infinite series

- 11.1. Integrals with an unbounded interval of integration — 273
- 11.2. Integrals of functions not defined in one point — 276
- 11.3. Calculation formulae — 280
- 11.4. Examples — 282
- 11.5. The Gamma function — 292
- 11.6. The relation between the convergence of an integral and the convergence of an infinite series — 294
- 11.7. Fourier series — 299
- 11.8. Applications and examples — 304
- Exercises on § 11 — 309

Supplement. Additional exercises — 311

Index — 328

Other titles in the Series in Pure and Applied Mathematics — 333

PREFACE TO THE SECOND ENGLISH EDITION

The English edition does not differ essentially from the Polish one. Among the more important supplements I should mention § 6.5 containing elementary information on the notation of mathematical logic. To this supplement I was inclined by the experience of many years. For many students (not for all, perhaps) the notation of definitions of certain notions by means of the logical symbols makes it easier to understand these notions (e.g. the notions of uniform continuity or uniform convergence). Besides that, this supplement is included in the book in such a manner that it can be omitted in reading the whole book. Among other changes introduced in the English text, I should mention the addition of a number of exercises and problems; in the second English edition, many of them have been collected in the Supplement. I am glad also to mention the simplification of certain proofs, and finally the removal of mistakes which were found in the primary text. This has been done with cooperation of the translator of this book, Dr. J. Musielak. I should like to express here my thanks for his work and for the accomplishment of the translation. I express also my gratitude to Professor A. Schinzel, thanking him for a number of corrections.

Warsaw, November 1968   *K. Kuratowski*

PREFACE TO THE POLISH EDITION

This book contains notes of lectures on differential and integral calculus, prepared for publication, which I have held for many years in the University of Warsaw.

The examples at the end of each section form a necessary complement of the course. On the other hand, the parts of the text marked by a star may be omitted at a first reading.

The first volume contains differential and integral calculus of functions of one variable. Function of two and more variables, partial derivatives and multiple integrals will be treated in the second volume.

Among the text-books which have been especially useful in preparing this book, I should mention those by Banach, Courant, Goursat, Hardy, Kowalewski, Mazurkiewicz, Sierpiński, Borsuk (a collection of exercises).

My best thanks are due to my colleagues, Professors Karol Borsuk and Władysław Nikliborc and Assistants Zygmunt Charzyński and Roman Sikorski for their numerous suggestions, hints and remarks in preparing my lecture for printing. I express my thanks also to the Department of Sciences of the Ministry of Education, the Editors of the Cooperative "Czytelnik", the Swedish Relief Committee and to "Monografie Matematyczne". I am very much obliged to them for their assistance in the publication of this book.

K. KURATOWSKI

Warszawa, October 1946

CHAPTER I

SEQUENCES AND SERIES

§ 1. INTRODUCTION

1.1. Various kinds of numbers

We shall assume that the notion of a real number (i.e. roughly speaking, a number having a finite or infinite decimal expansion) is known from the middle-school course ([1]). We shall recall the terminology and some properties of real numbers.

The numbers $1, 2, 3, \ldots$, are called *positive integers*. The number 0 belongs neither to the positive nor to the negative numbers. The fractional numbers, i. e. numbers of the form $\frac{p}{q}$, where p and q are integers and $q \neq 0$, are called also *rational numbers*. For the rational numbers the four arithmetical operations: addition, subtraction, multiplication and division (except division by 0) can always be carried out; to the symbol $\frac{p}{0}$ there corresponds no numerical value. Thus, the symbol ∞, which will be often used, does not mean any "infinite" number.

Geometrically, real numbers may be interpreted as points of a straight line upon which are fixed two points representing 0 and 1. These two points make it possible easily to construct all the rational points on this line, which is called the *numerical line*. For if we are given a rational number w, we take a segment of the length w

---

([1]) In §§ 1.8 and 1.9 of this section, we give two rigorous methods of introduction of real numbers in outline: an axiomatic method and another one, which reduces the notion of a real number to the notion of a rational number.

12   I. SEQUENCES AND SERIES

on the right side of 0 (or on the left side, if w is negative). The end of this segment is the point corresponding to the number w. It is well known that not all points of the straight line correspond to rational numbers, e.g. the number $\sqrt{2}$, i. e. the length of the diagonal of a square with the side equal to 1 is an irrational number.

Reasons of a geometric as well as of an algebraic nature suggest that we include in the notion of *number* all real numbers, i. e. irrational numbers as well as rational numbers. In this way the scope of algebraic operations increases: in the real domain, there exist also roots and logarithms of positive numbers.

It should also be noted that in the real domain not all operations can be carried out, e.g. there do not exist square roots of negative numbers and, consequently, equations of the second degree do not need to possess real roots. To avoid situations of this kind, complex numbers are introduced. However, we shall not consider complex numbers just now. When we use the word "number" here we shall always have in mind real numbers.

1.2. The principle of mathematical induction

Among the properties of positive integers we shall mention the following one which we shall often use. It is called the *principle of mathematical induction*. The principle may be stated as follows.

Let us assume that a property of positive integers is given, satisfying the following conditions:

 (i) the number 1 possesses this property,

 (ii) if the number n possesses this property, then the number $n+1$ possesses it, too.

The principle of induction states that, under these assumptions, any positive integer possesses the property stated.

The principle is in agreement with the following intuitive argument: if the number 1 possesses the considered property, then the second condition implies that the number 2 possesses this property; but then, again by the second condition, the number 3 possesses this property, similarly, this implies that the number 4 possesses the property etc. Obviously, we are not able to perform this reasoning to infinity. The principle of induction gives a mathematical formulation to our intuitive reasoning.

As an example of the application of the principle of induction, we shall quote the proof of the so-called *Bernoulli* [1] *inequality: for each positive integer n and for any real number* $a \geqslant -1$, *the formula*

(1) $$(1+a)^n \geqslant 1+na$$

holds.

We begin by noting that the expression $x \geqslant y$ means that x is greater than y or equal to y or that x is not less than y. Thus, $3 \geqslant 2$ holds as well as $3 \geqslant 3$.

In order to prove the Bernoulli inequality by using the principle of induction we must show:

(i) that the number 1 belongs to the set of numbers n possessing the property expressed by the inequality (1); this is obvious, for substituting the number 1 in place of n in the inequality (1) we obtain the inequality $1+a \geqslant 1+a$ which holds for every a;

(ii) that the formula (1) implies the formula obtained from (1) by substitution $n+1$ in place of n in (1), i.e. the formula

(1') $$(1+a)^{n+1} \geqslant 1+(n+1)a.$$

To prove this formula, let us multiply both sides of the inequality (1) by $1+a$. Since we have $1+a \geqslant 0$ by assumption, we obtain

$$(1+a)^{n+1} \geqslant (1+na)(1+a) = 1+(n+1)a+na^2.$$

---

[1] James Bernoulli, a Swiss mathematician in the first half of 18th century.

However, $na^2 \geq 0$; hence

$$1+(n+1)a+na^2 \geq 1+(n+1)a,$$

which yields the formula (1').

As a second example of the application of the principle of induction we shall give the proof of the following trigonometric relation

$$(2) \qquad \frac{1}{2}+\cos t+\cos 2t+\ldots+\cos nt = \frac{\sin\frac{2n+1}{2}t}{2\sin\frac{t}{2}},$$

satisfied by each integer $n \geq 0$ and by any real number t which is not an integral multiple of the number 2π.

In this case, we begin the induction with 0 instead of 1. When $n=0$, formula (2) is an identity $\frac{1}{2} = \frac{\sin\frac{1}{2}t}{2\sin\frac{1}{2}t} = \frac{1}{2}$. Thus, it remains to prove that formula (2) implies formula (2') obtained by substitution $n+1$ in place of n in (2).

Applying formula (2), we may write formula (2') in the following form:

$$(2') \qquad \frac{\sin\frac{2n+1}{2}t}{2\sin\frac{t}{2}}+\cos(n+1)t = \frac{\sin\frac{2n+3}{2}t}{2\sin\frac{t}{2}},$$

or (reducing to common denominator)

$$\sin\frac{2n+3}{2}t-\sin\frac{2n+1}{2}t = 2\sin\frac{t}{2}\cos(n+1)t.$$

But the last formula follows from the known formula for the difference of the sine. Hence we have proved that identity (2) implies (2'); thus, formula (2) holds for each $n \geq 0$, by the principle of induction.

1.3. The Newton (¹) binomial formula

First of all, we shall define the so-called *Newton* or *binomial coefficients* by the following formula:

$$(3) \qquad \binom{n}{k} = \frac{n(n-1)(n-2)\cdot\ldots\cdot(n-k+1)}{1\cdot 2\cdot 3\cdot\ldots\cdot k}.$$

Thus it is seen that the denominator is a product of successive positive integers from 1 to k, and the numerator is a product of k successive decreasing positive integers from n to $n-k+1$. Here we assume n and k to be positive integers and $n \geqslant k$.

We test immediately that $\binom{n}{1} = n$ and $\binom{n}{n} = 1$. We extend the definition of the symbol $\binom{n}{k}$ to the case where $k = 0$ by writing $\binom{n}{0} = 1$; we adopt this convention to simplify the technique of calculations.

Finally, let us note that $\binom{n}{k}$ is the number of combinations of n elements by k; we shall not use this fact here.

The following formula holds (for an arbitrary positive integer k):

$$(4) \qquad \binom{n}{k} + \binom{n}{k-1} = \binom{n+1}{k}.$$

Indeed,

$$\frac{n\cdot\ldots\cdot(n-k+2)(n-k+1)}{1\cdot\ldots\cdot(k-1)k} + \frac{n\cdot\ldots\cdot(n-k+2)}{1\cdot\ldots\cdot(k-1)}\cdot\frac{k}{k}$$
$$= \frac{n\cdot\ldots\cdot(n-k+2)(n-k+1+k)}{1\cdot\ldots\cdot(k-1)k} = \binom{n+1}{k}.$$

In the above reasoning we applied the formula (3); hence we had to assume that $k-1$ is a positive integer, i.e. that $k > 1$. However, if $k = 1$, the formula (4) may be checked directly, substituting $k = 1$.

(¹) Isaac Newton, 1642-1727, one of the most eminent scientists in history. One of the inventors of the differential and integral calculus.

Now, we shall prove the Newton binomial formula, i.e. the following formula:

(5) $(a+b)^n =$
$$= a^n + \binom{n}{1}a^{n-1}b + \binom{n}{2}a^{n-2}b^2 + \ldots + \binom{n}{k}a^{n-k}b^k + \ldots + \binom{n}{n}b^n.$$

It is easily seen that the right side is a sum of $n+1$ components; the first one is a^n and the last one b^n, since $\binom{n}{n} = 1$. This formula is obvious for $n = 1$. For $n = 2$ and $n = 3$ it is known from the middle-school course.

Since the formula (5) holds for $n = 1$, according to the principle of induction, assuming (5), it remains only to prove that the formula obtained by substitution $n+1$ in place of n in (5) is true. But $(a+b)^{n+1} = (a+b)^n(a+b) = (a+b)^n a + (a+b)^n b$, whence

$(a+b)^{n+1}$
$= a^{n+1} + \binom{n}{1}a^n b + \binom{n}{2}a^{n-1}b^2 + \ldots + \binom{n}{k}a^{n-k+1}b^k + \ldots + \binom{n}{n}ab^n +$
$\quad + a^n b + \binom{n}{1}a^{n-1}b^2 + \ldots + \binom{n}{k-1}a^{n-k+1}b^k + \ldots + \binom{n}{n-1}ab^n + \binom{n}{n}b^{n+1}$
$= a^{n+1} + \binom{n+1}{1}a^n b + \ldots + \binom{n+1}{k}a^{n-k+1}b^k + \ldots + \binom{n+1}{n}ab^n + b^{n+1},$

where we applied formula (4). The equation obtained in this way is just the equation obtained from (5) by substitution $n+1$ in place of n. Thus, the Newton binomial formula is proved.

We should note that the Newton binomial formula may be written in a neater way as follows:

(6) $$(a+b)^n = \sum_{k=0}^{n} \binom{n}{k} a^{n-k} b^k$$

(when $a \neq 0 \neq b$) ([1]).

([1]) When $a = 0$ or $b = 0$, formula (6) may be also applied with the restriction that the indeterminate symbol 0^0 (which is obtained for $k = n$ or $k = 0$) has to be replaced by 1.

1. INTRODUCTION

The symbol $\sum_{k=0}^{n}$ means that we must take the sum of $n+1$ terms which are obtained from the expression beyond this symbol by substituting successively all integers from 0 to n in place of k, and adding the terms so obtained.

Moreover, the binomial coefficients may be written as follows. Let us denote by $n!$ (read: factorial n) the product of all successive positive integers from 1 to n:

(7) $\quad n! = 1 \cdot 2 \cdot 3 \cdot \ldots \cdot n \quad$ (and let $0! = 1$ by convention).

It is easily seen that we have

(8) $$\binom{n}{k} = \frac{n!}{k!(n-k)!}$$

1.4*. Schwarz inequality

Applying the principle of induction, we shall prove the following inequality:

(9) $\quad (x_1 y_1 + \ldots + x_n y_n)^2 \leqslant (x_1^2 + \ldots + x_n^2)(y_1^2 + \ldots + y_n^2)$.

For $n = 1$ the formula (9) is obvious: $(x_1 y_1)^2 \leqslant x_1^2 y_1^2$. Thus, we have to prove this formula for $n+1$, assuming that it is true for n.

Now,

(10) $\quad (x_1 y_1 + \ldots + x_{n+1} y_{n+1})^2 = (x_1 y_1 + \ldots + x_n y_n)^2 +$
$+ 2(x_1 y_1 + \ldots + x_n y_n) x_{n+1} y_{n+1} + x_{n+1}^2 y_{n+1}^2$.

Moreover, $2ab \leqslant a^2 + b^2$, since $a^2 + b^2 - 2ab = (a-b)^2 \geqslant 0$; hence, writing $a = x_k y_{n+1}$ and $b = x_{n+1} y_k$, we have

$$2 x_k y_k x_{n+1} y_{n+1} \leqslant x_k^2 y_{n+1}^2 + x_{n+1}^2 y_k^2$$

and suming for $k = 1, 2, \ldots, n$, we obtain

$2(x_1 y_1 + \ldots + x_n y_n) x_{n+1} y_{n+1}$
$\qquad \leqslant (x_1^2 + \ldots + x_n^2) y_{n+1}^2 + (y_1^2 + \ldots + y_n^2) x_{n+1}^2$.

Applying (9) and (10) we get

$(x_1y_1 + \ldots + x_{n+1}y_{n+1})^2 \leqslant (x_1^2 + \ldots + x_n^2)(y_1^2 + \ldots + y_n^2) +$
$\quad + (x_1^2 + \ldots + x_n^2)y_{n+1}^2 + (y_1^2 + \ldots + y_n^2)x_{n+1}^2 + x_{n+1}^2 y_{n+1}^2$
$= (x_1^2 + \ldots + x_n^2)(y_1^2 + \ldots + y_{n+1}^2) + (y_1^2 + \ldots + y_{n+1}^2)x_{n+1}^2$
$= (x_1^2 + \ldots + x_{n+1}^2)(y_1^2 + \ldots + y_{n+1}^2).$

This yields

$(x_1y_1 + \ldots + x_{n+1}y_{n+1})^2 \leqslant (x_1^2 + \ldots + x_{n+1}^2)(y_1^2 + \ldots + y_{n+1}^2).$

But this is just the formula obtained from (9) by substituting $n+1$ in place of n. Thus, the proof is finished.

1.5. The principle of continuity (Dedekind ([1]))

Among the properties of real numbers, we shall mention the so-called *continuity principle*. This principle states that if we divide the set of all real numbers into two sets A and B such that any number belonging to the set A is less than any number belonging to the set B, then there are two possibilities: either there exists a greatest number of the set A or there exists a least number of the set B (it being assumed that both the sets are non-empty). In other words, if we divide the straight line in two parts A and B in such a way that any point of the part A lies on the left side of any point of the part B, then either there exists a last point in the part A or there exists the first point in the part B. No "gap" may be found in this "cut" which we have performed. This implies the continuity of the set of real numbers distinguishing this set from the set of rational numbers, where such gaps exist; e.g. dividing the set of rational numbers in two parts such that all rational numbers $< \sqrt{2}$ belong to the first part and all other rational numbers, i.e. numbers $> \sqrt{2}$, belong to the second part, it is easily seen that any number of the first part is less

---

[1] Richard Dedekind, a German mathematician in the second half of the 19th century.

than any number of the second part, but there does not exist any rational number which would be the greatest in the first part or the least in the second one (this follows from the fact that the irrational number $\sqrt{2}$ may be approximated arbitrarily accurate by the rational numbers from below and from above, e.g. by the decimal expansions).

1.6. The absolute value of a number

The absolute value of a number a, i.e. $|a|$, is defined by the conditions: if $a \geqslant 0$, then $|a| = a$; if $a < 0$, then $|a| = -a$. The following formulae, known from elementary mathematics, hold:

(11) $\qquad |a| = |-a|$,

(12) $\qquad -|a| \leqslant a \leqslant |a|$,

(13) $\qquad |a+b| \leqslant |a|+|b|$,

(14) $\qquad |a|-|b| \leqslant |a-b| \leqslant |a|+|b|$,

(15) $\qquad |ab| = |a|\cdot|b|$,

(16) *the inequalities* $|a| \leqslant c$ *and* $|b| \leqslant d$ *imply* $|a+b| \leqslant c+d$;

the proofs will be left to the reader.

The last formula (16) is the formula for the addition of inequalities "under the sign of absolute value" which we shall often apply. It follows from the formula (13), since $|a+b| \leqslant |a|+|b| \leqslant c+d$. This formula remains true, if the sign $\leqslant$ is replaced by $<$.

Finally, let us note that

(17) *the inequality* $|a| < b$ *is equivalent to the double inequality* $-b < a < b$ *and hence to the collection of two inequalities*: $a < b$ *and* $-a < b$.

1.7. Bounded sets. The upper bound and the lower bound of a set

We say that a set of real numbers Z is *bounded*, if two numbers m and M exist such that every number x belonging to the set Z satisfies the double inequality $m \leqslant x \leqslant M$. Assuming only that there exists a number M

satisfying the inequality $M \geqslant x$ for every x belonging to the set Z, we call the set Z *bounded above*. Similarly, a set Z is called *bounded below* if a number m exists satisfying the above condition: $m \leqslant x$.

The geometrical meaning of these ideas is the following. If a set is bounded it means that it is contained in a certain segment on the numerical line. A set is bounded above or below, when it is contained in an infinite radius directed to the left or to the right, respectively.

Applying the continuity principle we shall prove the following theorem:

If a non-empty set Z is bounded above, then among the numbers M, satisfying the inequality $M \geqslant x$ for any x belonging to Z, there exists a least one. This number will be called the upper bound of the set Z.

Similarly, *if a set Z is bounded below, then among the numbers m, satisfying the inequality $m \leqslant x$ for any x belonging to Z, there exists a greatest one, which is called the lower bound of the set Z.*

Proof. Let the set Z be bounded above. Let us divide the set of all real numbers in two classes as follows. To the second class there will belong all numbers M satisfying the inequality $M \geqslant x$ for any x belonging to Z. To the first class there will belong all other real numbers; that means that a number a belongs to the first class when in the set Z there exists a number greater than a. Such a division of the set of real numbers in two classes is called a *cut*, i. e. any number belonging to the first class is less than any number belonging to the second one. Indeed, supposing a number M of the second class to be less than a number a of the first class and knowing that a number x exists in the set Z such that $a < x$ we should have $M < x$; but this contradicts the definition of the second class.

Moreover, let us note that both classes are non-empty. Indeed, if a number z belongs to the set Z (and such

1. INTRODUCTION 21

a number exists, the set Z being non-empty by assumption), then the number $z-1$ belongs to the first class. The second class is also non-empty, since the set Z is bounded above.

According to the continuity principle there exists either a greatest number in the first class or a least number in the second class. However, the first eventuality is not possible. Namely, if a belongs to the first class and $a < x$ (where x belongs to Z), then denoting by a' any number between a and x, e. g. $a' = \dfrac{a+x}{2}$, we have also $a' < x$, but this means that a' belongs to the first class, too. Thus, to any number a of the first class we may find a number a' greater than a in this class. This means that in the first class a greatest number does not exist. Hence, there exists in the second class a least number, i. e. the least number among the numbers M satisfying the inequality $M \geqslant x$ for every x belonging to the set Z. Hence the theorem is proved.

The proof of existence of a lower bound is completely analogous.

Let us note that the upper bound and the lower bound of a set Z do not necessarily belong to this set. E. g. the bounds of the open interval ([1]) $a < x < b$ are the numbers a and b which do not belong to this interval.

1.8*. The axiomatic treatment of real numbers

The notion of a real number which we have assumed to be known from the middle-school course may be introduced in an axiomatic way as follows.

We assume that in the set of real numbers two operations can be performed: *addition* $x+y$ and *multiplication* xy. These operations satisfy the laws of *com-*

([1]) By an open interval we understand the set of numbers x such that $a < x < b$; by a closed interval we understand the open interval together with ends, i.e. the set of x such that $a \leqslant x \leqslant b$.

mutativity and *associativity*:
$$x+y = y+x, \quad xy = yx,$$
$$(x+y)+z = x+(y+z), \quad (xy)z = x(yz).$$

Moreover, the multiplication is *distributive* with respect to the addition:
$$x(y+z) = xy+xz.$$

Two (different) numbers 0 and 1 are the *moduli* of addition and multiplication respectively, i.e.
$$x+0 = x, \quad x \cdot 1 = x.$$

Further, we assume that in the set of real numbers subtraction and division are always possible, except division by 0. In other words, we assume that to any pair of numbers x and y a number z (called the *difference* $x-y$) exists such that
$$x = y+z$$
and, in the case where $y \neq 0$, a number w (called the *quotient* $x:y$) exists such that
$$x = yw.$$

Besides the above axioms concerning the operations, we take following axioms concerning the order relation $x < y$. We assume that any two distinct real numbers x and y are in this relation in one or another direction, i.e. either $x < y$ or $y < x$. This relation is *transitive*, i.e.
$$\text{the conditions } x < y \text{ and } y < z \text{ imply } x < z$$
and *asymmetric*, i.e.
$$\text{if } x < y, \text{ then the relation } y < x \text{ does not hold.}$$

The order relation is connected with the basic operations by the following axiom:
$$\text{if } y < z, \text{ then } x+y < x+z \text{ and if, moreover, } 0 < x,$$
then $xy < xz$.

Finally, the last axiom in the theory of real numbers which we are accepting is the *Dedekind continuity principle* formulated in § 1.5.

All arithmetic and algebraic theorems known from courses in the middle-school may be deduced from the above axioms.

1.9*. Real numbers as sets of rational numbers [1]

The notion of a real number may be defined on the basis of the theory of rational numbers as follows.

Real numbers may be considered to be identical with sets of rational numbers R, satisfying the following conditions:

(i) the set R does not contain a greatest number, i.e. for any number belonging to the set R there exists a greater number in R;

(ii) if a number x belongs to R, then any rational number less than x belongs to R;

(iii) the set R is non-empty and is not equal to the set of all rational numbers.

For real numbers defined in this way we define first of all the relation of order. Namely, we write $R < R'$, when the set R is a part of the set R' (different from R'); or, when the set R' contains numbers which do not belong to R (it is easily seen that for any two sets satisfying the conditions (i)–(iii) always one is contained in the other). Both these definitions are equivalent.

We easily find that the order relation defined in this way satisfies the axioms given in § 1.8, i.e. it is a transitive, asymmetric relation which holds for any pair of different sets R and R' in one or other direction.

Now we define the addition of real numbers, i.e. the addition of sets R and R' (satisfying the conditions

---

[1] We give here an outline of the so-called Dedekind theory of real numbers.

(i)–(iii)). Namely, by $R+R'$ we denote the set of all numbers which are sums of two numbers, the first of which belongs to R and the second one belongs to R'. It is easily seen that the set $R+R'$ satisfies the conditions (i)–(iii). Moreover, the axioms concerning the addition (commutativity, associativity etc.) are satisfied. The real number 0 is defined as the set of all negative rational numbers. Similarly, the real number 1 is the set of all rational numbers less than the rational-one. In general, if r is a rational number, then we understand by "the real number r" the set of all rational numbers less than the rational number r (in practice, we identify the rational number r and the real number r). $-R$ means the set of all rational numbers of the form $-x$, where x takes all values which do not belong to R; however, if the set of numbers not belonging to R contains a least number r, then we do not include the number $-r$ in the set $-R$ (that is the case, when R is "the real number r"). One may prove that $R+(-R) = 0$.

The multiplication of real numbers will be defined in the following way: if $R \geqslant 0$ and $R' \geqslant 0$, then RR' is the set containing all negative rational numbers and numbers of the form rr', where r is a non-negative number belonging to R and, similarly, r' is a non-negative number belonging to R'. Moreover, we assume

$$(-R)(-R') = RR', \quad (-R)R' = -(RR') = R(-R').$$

It may easily be proved that all the axioms concerning the multiplication are then satisfied.

Finally, the continuity axiom is satisfied. Indeed, let A, B denote a cut in the domain of real numbers. We denote by R the set of rational numbers with the property that r belongs to R if and only if it belongs to one of the real numbers belonging to the class A.

It is proved that the above defined set R satisfies the conditions (i)–(iii), so that it is a real number. Then we prove that R "lies on the cut A, B", i.e. that it is either the greatest number in the set A or the least number in the set B.

Thus, we see that, if we understand by a real number a set of rational numbers satisfying the conditions (i)–(iii), then all axioms of the theory of real numbers formulated in § 1.8 are satisfied.

Let us add that the set of all rational numbers satisfies all these axioms with the exception of one axiom, namely the continuity axiom.

Hence it follows that the continuity axiom is independent of the other axioms of the theory of real numbers; for if it were a consequence of the others (i.e. a theorem and not an axiom), then it would be satisfied in every system in which the other axioms are satisfied.

Exercises on § 1

1. Prove that for each positive integer n and for any real $a \geqslant -1$ the following formula holds (being a generalization of the Bernoulli formula (1)):

$$(1+a)^n \geqslant 1 + na + \frac{n(n-1)}{2} a^2 + \frac{n(n-1)(n-2)}{6} a^3.$$

Hint: proof by the mathematical induction.

2. Prove the identities:

$$1 + 2 + \ldots + n = \frac{n(n+1)}{2},$$

$$1^2 + 2^2 + \ldots + n^2 = \frac{n(n+1)(2n+1)}{6}.$$

Hint: in the proof of the second equality one may apply the identity:

$$(k+1)^3 - k^3 = 3k^2 + 3k + 1.$$

3. Similarly: find the formula for the sum

$$1^3 + 2^3 + \ldots + n^3.$$

4. Deduce the formulae:

$$\binom{n}{0} + \binom{n}{1} + \ldots + \binom{n}{n} = 2^n,$$

$$\binom{n}{0} - \binom{n}{1} + \binom{n}{2} - \ldots \pm \binom{n}{n} = 0.$$

Hint: apply the Newton binomial formula.

5. Applying the continuity principle prove that $\sqrt[n]{a}$ exists for each positive integer n and for any positive a.

6. Prove that $||a|-|b|| \leqslant |a-b|$. Prove that the equation $|a+b| = |a|+|b|$ holds, if and only if $ab \geqslant 0$.

7. Denote by A and G the arithmetic and the geometric mean of the numbers $a_1, a_2, \ldots, a_n$, respectively, i.e.

$$A = \frac{a_1 + a_2 + \ldots + a_n}{n}, \qquad G = \sqrt[n]{a_1 \cdot a_2 \cdot \ldots \cdot a_n}.$$

Prove that if the numbers $a_1, a_2, \ldots, a_n$ are positive, then $G \leqslant A$.

Hint: precede the proof by the remark that if $a_1 < A < a_2$, then the arithmetic mean of the numbers $A, a_1 + a_2 - A, a_3, a_4, \ldots, a_n$ is A and the geometric mean of these numbers is $> G$.

8. Denoting by H the harmonic mean of the above numbers $a_1, a_2, \ldots, a_n$, i.e. writing

$$H = n : \left(\frac{1}{a_1} + \frac{1}{a_2} + \ldots + \frac{1}{a_n} \right),$$

prove that $H \leqslant G$.

§ 2. INFINITE SEQUENCES

2.1. Definition and examples

If a real number corresponds to each positive integer, then we say that an infinite sequence is defined.

For instance, positive even numbers constitute an infinite sequence $2, 4, 6, \ldots, 2n, \ldots$; namely, to the number 1 there corresponds the number 2, to the number 2, the number 4, to the number 3, the number 6, and, generally, to the positive integer n there corresponds the number $2n$.

Usually, we write an infinite sequence in the form $a_1, a_2, \ldots, a_n, \ldots$ or $\{a_n\}$. The numbers appearing in the sequence are called the *terms* of the sequence; the expression a_n is called the *general term* of the sequence. Thus, for the sequence of positive even numbers the general term is $2n$; for the sequence of odd numbers the general term is $2n-1$.

In an infinite sequence some terms may be repeated. E.g. if the number 0 corresponds to the odd positive integers and the number 1 to the even positive integers, then the sequence $0, 1, 0, 1, \ldots$ is obtained. This sequence consists of two alternating terms.

In particular, all terms of a sequence may be identical: $c, c, c, \ldots$

As examples of sequences we shall mention the arithmetic and the geometric progressions. The arithmetic progression is a sequence of the form $c, c+d, c+2d, \ldots$, i.e. a sequence with the general term $c+(n-1)d$. Similarly, the general term of a geometric progression is $a_n = cq^{n-1}$.

A *definition of a sequence by induction* (i.e. a *recurrence* definition) is often used. Here, we define directly the term a_1, but a_n is defined by means of the earlier terms, e.g., write $a_1 = 1$, $a_n = 2^{a_{n-1}}$. The first terms of this sequence are: $1, 2, 4, 16, 2^{16}, \ldots$

Finally, let us note that it may not always be possible to write the general term by means of a formula. For example the prime numbers constitute, as is well known, an infinite sequence but we do not know any mathematical formula defining the n-th prime number.

Geometrically, we interprete an infinite sequence as a set of points on the plane (on which the axis of abscissae and the axis of ordinates are given; cf. Fig. 1, p. 30).

Namely, the sequence $a_1, a_2, ..., a_n, ...$ is the set of points $(1, a_1), (2, a_2), ..., (n, a_n), ...$ For example the set of positive integers is the intersection of the straight line $y = x$ and of the pencil of straight lines parallel to the Y-axis passing through the positive integer points on the X-axis. Similarly, the sequence with a constant term c is the intersection of the above pencil and of the straight line parallel to the X-axis. If we take on the straight lines of this pencil points with the abscissae 0 and 1, alternately, we obtain the oscillating sequence $0, 1, 0, 1, ...$

As may by seen from the above remarks, an infinite sequence and the set of the terms of this sequence are two distinct notions: a sequence is a set of points of the plane and the set of terms of a sequence is a certain set of real numbers, i.e. from the geometrical point of view, a set on the straight line. Two different sequences may consist of the same terms: the above defined oscillating sequence $0, 1, 0, 1, ...$ and the sequence $0, 1, 1, 1, ...$ (having at the first place 0 and at all other places 1) have the same terms; however, they are sequences with fundamentally different properties.

A sequence is called *increasing*, if $a_1 < a_2 < a_3 ...$, in general: if $a_n < a_{n+1}$. Similarly, a sequence is called *decreasing*, if $a_n > a_{n+1}$. If we replace the sign $<$ by $\leqslant$ (or $>$ by $\geqslant$), we obtain increasing (or decreasing) sequences in the wider sense, called also *non-decreasing* (or *non-increasing*) sequences. Increasing and decreasing sequences in the wider sense are generally called *monotone sequences*.

The sequence of even numbers is an increasing sequence, the sequence $1, 1, 2, 2, 3, 3, ...$ is an increasing sequence in the wider sense. The oscillating sequence $0, 1, 0, 1, ...$ is neither increasing nor decreasing.

2.2. The notion of limit

A number g is called the *limit* of the infinite sequence $a_1, a_2, ..., a_n, ...$ if to any positive number ε a number k

exists such that for $n > k$, the inequality
(1) $$|a_n - g| < \varepsilon,$$
holds, i.e. a_n lies between $g - \varepsilon$ and $g + \varepsilon$.
We denote the limit of a sequence by the symbol
(2) $$g = \lim_{n \to \infty} a_n.$$

Let us take the sequence with the general term $a_n = \dfrac{n+1}{n}$ as an example of a sequence with a limit, i.e. the sequence $2, 1\frac{1}{2}, 1\frac{1}{3}, ..., 1\frac{1}{n}, ...$ We shall prove that 1 is the limit of this sequence.

Let a number $\varepsilon > 0$ be given. We have to choose k in such way that for $n > k$, the inequality $\left|\dfrac{n+1}{n} - 1\right| < \varepsilon$ would hold, i.e. that $\dfrac{1}{n} < \varepsilon$. Now, let k be a positive integer greater than $\dfrac{1}{\varepsilon}$. Then we have $\dfrac{1}{k} < \varepsilon$; thus, the double inequality $\dfrac{1}{n} < \dfrac{1}{k} < \varepsilon$ holds for $n > k$, whence $\dfrac{1}{n} < \varepsilon$, as we set out to prove.

Let us note that we have proved simultaneously that
$$\lim_{n \to \infty} \frac{1}{n} = 0.$$

The number k chosen to the given ε depends on ε. Mostly, if we diminish ε, k has to increase (this fact is easy to observe in the above example).

To express this more figuratively, the condition for g to be the limit of a sequence $a_1, a_2, ...$ may be formulated as follows: for any $\varepsilon > 0$, the inequality (1) is satisfied for all sufficiently large values of n.

A sequence having a limit is called *convergent*. A *divergent* sequence is a sequence which has no limit. As an example of a divergent sequence, let us take the sequence of all positive integers. Indeed, let us suppose that g is the limit of the sequence of positive integers. Writing 1 in place of ε we have, by (1), the inequality $|n-g| < 1$ for sufficiently large n; thus $n < g+1$. Here is a contradiction, since just the reverse inequality $n > g+1$ holds for sufficiently large n. This contradiction shows that our supposition that the sequence of positive integers has a limit was false. Thus, it is a divergent sequence.

Similarly, one might prove that the oscillating sequence $0, 1, 0, 1, \ldots$ is divergent.

Further,

(3) $$\lim_{n=\infty} c = c,$$

i.e. the sequence with constant term c has c as the limit.

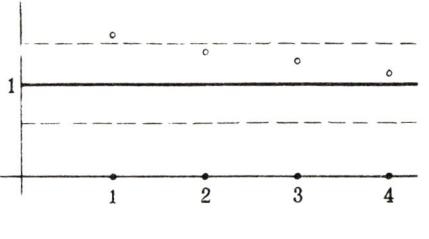

Fig. 1

Namely, we have always $a_n - c = 0$. Thus, the inequality (1) holds for all n.

Geometrically, the notion of limit may be interpreted as follows. Let a sequence $a_1, a_2, \ldots$ be given (e.g. the sequence with the general term $\frac{n+1}{n}$). Let us imagine the geometrical interpretation of this sequence on the plane according to § 2.1. Let us draw a straight line parallel to the X-axis through the point g of the Y-axis (e.g. $g = 1$). Moreover, let us draw two straight lines parallel to the last one on two different sides of it but in the same distant. The condition that the equality (2) holds means that for sufficiently large n (i. e. for n larger than a certain k), all terms of our sequence lie in the strip determined by the two above-defined parallel straight lines.

The above geometrical condition may be easily obtained from the analytical condition formulated in the definition of limit denoting the breadth of the above strip by 2ε.

In other words, the condition for the convergence of a sequence to g may be formulated as follows: every interval with the centre g contains all elements of the sequence beginning with a certain index.

In the above examples we have seen that a sequence may have no limit (if it is divergent); the question arises whether if a limit exists (thus, if the sequence is convergent), then it is unique. We shall show that it is really the case.

For let us assume a sequence $a_1, a_2, \ldots$ to have two different limits g and g'. Then we have $|g-g'| > 0$. Let $\varepsilon = \frac{1}{2}|g-g'|$. According to the definition of the limit, two numbers k and k' exist such that for $n > k$ the inequality

(4) $$|a_n - g| < \varepsilon,$$

and for $n > k'$ the inequality

(5) $$|a_n - g'| < \varepsilon$$

hold.

Thus, denoting by m the greater among the two numbers k and k', the inequalities (4) and (5) hold for $n > m$, simultaneously. Let us add both sides of these inequalities under the sign of absolute value (cf. § 1 (15)), after changing the sign under the absolute value in the formula (4). We obtain $|g-g'| < 2\varepsilon$, but $2\varepsilon = |g-g'|$. Thus, we get a contradiction.

Remark. It is easily seen that the notion of limit is not changed, if we replace in the definition of limit the sign $>$ by $\geqslant$ or $<$ by $\leqslant$.

2.3. Bounded sequences

A sequence $a_1, a_2, \ldots$ is called *bounded* if the set of its terms is bounded, i. e. if there exists a number M such that the inequality $|a_n| < M$ holds for all values of n. In other words, the inequality $-M < a_n < M$ is satisfied for all terms of the sequence.

The geometrical interpretation of this condition is that the whole sequence (treated as a set on the plane) is contained between two straight lines $y = M$ and $y = -M$.

We say also that a sequence is *bounded above* if a number M exists such that $a_n < M$ for all n, i.e. that the sequence lies below the straight line $y = M$. The notion of a sequence *bounded below* can be defined analogously. Clearly, a sequence which is bounded above and below is simply a bounded sequence.

EXAMPLES. The sequence $0, 1, 0, 1, \ldots$ is bounded. The sequence of positive integers is unbounded, though it is bounded below. The sequence $1, -1, 2, -2, 3, -3, \ldots$ is neither bounded above nor bounded below.

THEOREM. *Every convergent sequence is bounded.*

Indeed, let us assume that the equation (2) holds and let us substitute the value 1 for ε. Hence a number k exists such that we have $|a_n - g| < 1$ for $n > k$. Since $|a_n| - |g| \leqslant |a_n - g|$ (cf. § 1 (13)), we have $|a_n| < |g| + 1$. Let us denote by M a number greater than any among the following $k+1$ numbers: $|a_1|, |a_2|, \ldots, |a_k|, |g|+1$. Since the last one is greater than $|a_{k+1}|, |a_{k+2}|$ etc., we get $M > |a_n|$ for each n. Thus, the sequence is bounded.

2.4. Operations on sequences

THEOREM. *Assuming the sequences $a_1, a_2, \ldots$ and $b_1, b_2, \ldots$ to be convergent, the following four formulae hold* ([1]):

(6) $\quad \lim(a_n + b_n) = \lim a_n + \lim b_n$,
(7) $\quad \lim(a_n - b_n) = \lim a_n - \lim b_n$,

---

[1] To simplify the symbolism we shall often omit the equality $n = \infty$ under the sign of lim.

(8) $\lim(a_n \cdot b_n) = \lim a_n \cdot \lim b_n$,

(9) $\lim \dfrac{a_n}{b_n} = \dfrac{\lim a_n}{\lim b_n}$ (*when* $\lim b_n \neq 0$).

This means that under our assumptions the limit of the sum exists and is equal to the sum of limits, the limit of the difference exists and is equal to the difference of limits etc.

Proof. Let us write $\lim a_n = g$ and $\lim b_n = h$. A number $\varepsilon > 0$ let be given. Hence a number k exists such that the inequalities $|a_n - g| < \varepsilon/2$ and $|b_n - h| < \varepsilon/2$ hold for $n > k$. We add both these inequalities under the sign of absolute value. We obtain

$$|(a_n + b_n) - (g + h)| < \varepsilon.$$

This means that the sequence with general term $c_n = a_n + b_n$ is convergent to the limit $g + h$. Thus, we have proved the formula (6).

In particular, if b_n takes a constant value: $b_n = c$, formulae (6) and (3) imply:

(10) $\lim(a_n + c) = c + \lim a_n$.

Now, we shall prove the formula (8). We have to "estimate" the difference $|a_n b_n - gh|$. To be able to apply the convergence of the sequences $a_1, a_2, \ldots$ and $b_1, b_2, \ldots$ we transform this difference as follows:

$$a_n b_n - gh = a_n b_n - a_n h + a_n h - gh = a_n(b_n - h) + h(a_n - g).$$

Since the sequence $a_1, a_2, \ldots$ is convergent, it is bounded and so a number M exists such that $|a_n| < M$. Applying to the last equation the formulae for the absolute value of a sum and of a product we get:

$$|a_n b_n - gh| \leqslant |a_n(b_n - h)| + |h(a_n - g)|$$
$$\leqslant M \cdot |b_n - h| + |h| \cdot |a_n - g|.$$

Now, let us take a number $\eta > 0$ independently of ε. Hence a number k exists such that we have $|a_n - g| < \eta$

and $|b_n-h| < \eta$ for $n > k$. Thus,
$$|a_n b_n - gh| < M\eta + |h|\eta = (M+|h|)\eta.$$

Till now we have not assumed anything about the positive number η. Let us now assume that $\eta = \varepsilon/(M+|h|)$. So we conclude that the inequality $|a_n b_n - gh| < \varepsilon$ holds for $n > k$. Thus, we have proved the formula (8).

In particular, if we write $b_n = c$ we get

(11) $\qquad \lim(c \cdot a_n) = c \lim a_n,$
(12) $\qquad \lim(-a_n) = -\lim a_n,$

where the formula (12) follows from (11) by the substitution $c = -1$.

Formulae (6) and (12) imply the formula (7), for
$$\lim(a_n - b_n) = \lim[a_n + (-b_n)]$$
$$= \lim a_n + \lim(-b_n) = \lim a_n - \lim b_n.$$

Before proceeding to the proof of the formula (9), we shall prove the following special case of this formula:

(13) $\qquad \lim \dfrac{1}{b_n} = \dfrac{1}{\lim b_n} \qquad$ (when $\lim b_n \neq 0$).

First, we note that for sufficiently large n the inequality $b_n \neq 0$ holds. We shall prove an even stronger statement: we have $|b_n| > \frac{1}{2}|h|$ for sufficiently large n. Indeed, since $\frac{1}{2}|h| > 0$, a number k exists such that $|b_n - h| < \frac{1}{2}|h|$ for $n > k$. Hence,
$$|h| - |b_n| \leqslant |h - b_n| < \tfrac{1}{2}|h| \quad \text{and thus} \quad |b_n| > \tfrac{1}{2}|h|.$$
To prove the formula (13), the difference
$$\left|\frac{1}{b_n} - \frac{1}{h}\right| = \left|\frac{h-b_n}{h \cdot b_n}\right| = \frac{|h-b_n|}{|h| \cdot |b_n|}$$
has to be estimated.

But for sufficiently large n we have $|h-b_n| < \eta$ and $|b_n| > \frac{1}{2}|h|$, i.e. $1/|b_n| < 2/|h|$. Thus,
$$\left|\frac{1}{b_n} - \frac{1}{h}\right| < \frac{2\eta}{h^2}.$$

Writing $\eta = \frac{1}{2}\varepsilon h^2$, we get

$$\left|\frac{1}{b_n} - \frac{1}{h}\right| < \varepsilon,$$

whence the formula (13) follows.

The formula (9) follows from (8) and (13):

$$\lim \frac{a_n}{b_n} = \lim a_n \cdot \frac{1}{b_n} = \lim a_n \cdot \lim \frac{1}{b_n} = \frac{\lim a_n}{\lim b_n}.$$

Remarks. (α) We have assumed that the sequences $\{a_n\}$ and $\{b_n\}$ are convergent. This assumption is essential, for it may happen that the sequence $\{a_n + b_n\}$ is convergent, although both the sequences $\{a_n\}$ and $\{b_n\}$ are divergent; then the formula (6) cannot be applied. As an example one can take: $a_n = n$, $b_n = -n$.

However, if the sequence $\{a_n + b_n\}$ and one of the two sequences, e. g. the sequence $\{a_n\}$ are convergent, then the second one is also convergent. For $b_n = (a_n + b_n) - a_n$, and so the sequence $\{b_n\}$ is convergent as a difference of two convergent sequences.

Analogous remarks may be applied to the formulae (7)–(9).

(β) In the definition of a sequence we have assumed that the enumeration of the elements begins with 1. It is convenient to generalize this definition assuming that the enumeration begins with an arbitrary positive integer (and even with an arbitrary integer), e.g. with 2, 3 or another positive integer. So is e.g. in the proof of the formula (13). We have proved that $b_n \neq 0$ beginning with a certain k. Thus, the sequence $\frac{1}{b_n}$ is defined just beginning with this k (for if $b_n = 0$, then $\frac{1}{b_n}$ does not mean any number).

This remark is connected with the following property of sequences, easy to prove: *the change of a finite number of terms of a sequence has influence neither on the con-*

vergence of the sequence nor on its limit. The same can be said of the addition or omission of a finite number of terms in a sequence.

EXAMPLES. Find $\lim_{n=\infty} \frac{6n+2}{7n-3}$. In this case we cannot apply the formula (9) directly, since neither the numerator nor the denominator are convergent, as n tends to ∞. However, we can transform the general term of the sequence $a_n = \frac{6n+2}{7n-3}$ to become a quotient of two expressions each of which has a limit. For this purpose it is sufficient to divide the numerator and the denominator by n. Then we obtain $a_n = \frac{6+2/n}{7-3/n}$ and we may apply the formula (9). Since

$$\lim_{n=\infty} \left(6 + \frac{2}{n}\right) = 6 \quad \text{and} \quad \lim_{n=\infty} \left(7 - \frac{3}{n}\right) = 7,$$

we have

$$\lim_{n=\infty} a_n = \frac{6}{7}.$$

Similarly,

$$\lim_{n=\infty} \frac{n+1}{n^2+1} = \lim_{n=\infty} \frac{\frac{1}{n}+\frac{1}{n^2}}{1+\frac{1}{n^2}} = 0,$$

for

$$\lim_{n=\infty} \frac{1}{n} = 0 = \lim_{n=\infty} \frac{1}{n^2}.$$

2.5. Further properties of the limit

Suppose that a sequence $\{a_n\}$ is convergent. Then the sequence $\{|a_n|\}$ is convergent, too, and

(14) $\qquad \lim |a_n| = |\lim a_n|$.

Let $\lim a_n = g$. Then we have $|a_n - g| < \varepsilon$ for sufficiently large n. Thus,

$$|a_n| - |g| \leqslant |a_n - g| < \varepsilon \quad \text{and} \quad |g| - |a_n| \leqslant |g - a_n| < \varepsilon,$$

whence (cf. § 1 (17)): $||a_n|-|g|| < \varepsilon$. Thus, the formula (14) is proved.

Assuming the sequences $\{a_n\}$ and $\{b_n\}$ to be convergent, the following relation holds:

(15) *the inequality* $a_n \leqslant b_n$ *implies* $\lim a_n \leqslant \lim b_n$.

In particular, if the sequence $\{c_n\}$ is convergent, then

(16) *the condition* $c_n \geqslant 0$ *implies* $\lim c_n \geqslant 0$.

We shall prove first the last formula. Let $\lim c_n = h$, and let us further assume that $h < 0$, i.e. that $-h > 0$. Then we have $|c_n - h| < -h$ for sufficiently large n, and hence $c_n - h < -h$, whence $c_n < 0$, which contradicts our assumption.

Applying the formula (16) we shall prove now the formula (15).

We put $b_n - a_n = c_n$. Since $a_n \leqslant b_n$, we have $c_n \geqslant 0$ and thus, in the limit, $\lim c_n \geqslant 0$. Moreover, (7) implies:

$$\lim c_n = \lim b_n - \lim a_n,$$

hence

$$\lim b_n - \lim a_n \geqslant 0, \quad \text{i.e.} \quad \lim a_n \leqslant \lim b_n.$$

Remark. In the formulation of the relation (16), the inequality $\geqslant$ cannot be replaced by $>$ (similarly, in (15) one cannot replace $\leqslant$ by $<$). For example the sequence $c_n = 1/n$ satisfies the inequality $c_n > 0$, but $\lim c_n = 0$.

Thus we see that the relations $\leqslant$ and $\geqslant$ "remain true in the limit", but the relations $<$ and $>$ do not possess this property.

We next prove the *formula of the double inequality*:

(17) *If* $a_n \leqslant c_n \leqslant b_n$ *and if* $\lim a_n = \lim b_n$, *then the sequence* $\{c_n\}$ *is convergent and* $\lim c_n = \lim a_n = \lim b_n$.

Suppose that $\lim a_n = g = \lim b_n$ and let $\varepsilon > 0$. Then we have $|a_n - g| < \varepsilon$ and $|b_n - g| < \varepsilon$ for sufficiently large n. According to the assumption,

$$a_n - g \leqslant c_n - g \leqslant b_n - g,$$
$$\text{and} \quad -\varepsilon < a_n - g \text{ and } b_n - g < \varepsilon;$$

hence $-\varepsilon < c_n - g < \varepsilon$, i. e. $|c_n - g| < \varepsilon$, whence $\lim c_n = g$.

(18)   *If* $\lim |a_n| = 0$, *then the sequence* $\{a_n\}$ *is convergent and* $\lim a_n = 0$.

For, we have then $-|a_n| \leqslant a_n \leqslant |a_n|$ and
$$\lim(-|a_n|) = 0 = \lim |a_n|.$$

2.6. Subsequences

Let a sequence $a_1, a_2, ..., a_n, ...$ and an increasing sequence of positive integers $m_1, m_2, ..., m_n, ...$ be given. The sequence
$$b_1 = a_{m_1}, \quad b_2 = a_{m_2}, \quad ..., \quad b_n = a_{m_n}, \quad ...$$
is called a *subsequence of the sequence* $a_1, a_2, ..., a_n, ...$

For example the sequence $a_2, a_4, ..., a_{2n}, ...$ is a subsequence of the sequence $a_1, a_2, ...$ Yet, the sequence $a_1, a_1, a_2, a_2, ...$ is not a subsequence of this sequence, since in this case the indices do not form an increasing sequence.

We have the general formula
(19) $m_n \geqslant n$.

This is obvious for $n = 1$, i.e. we have $m_1 \geqslant 1$ (since m_1 is a positive integer). Applying the principle of induction, let us assume that the formula (19) holds for a given n. Then we have $m_{n+1} > m_n \geqslant n$, whence $m_{n+1} \geqslant n+1$. So we have obtained the formula (19) for $n+1$. Thus, the formula (19) is true for each n.

According to our definition, every sequence is its own subsequence. We can say in general that every subsequence is obtained from the sequence by omitting a number of elements in this sequence (this number may be finite, infinite or zero). Hence it follows also that a subsequence $\{a_{m_{k_n}}\}$ of a subsequence $\{a_{m_n}\}$ is a subsequence of the sequence $\{a_n\}$.

THEOREM 1. *A subsequence of a convergent sequence is convergent to the same limit. In other words,*

(20)   *if* $\lim\limits_{n=\infty} a_n = g$ *and if* $m_1 < m_2 < ...$, *then* $\lim\limits_{n=\infty} a_{m_n} = g$.

Let a number $\varepsilon > 0$ be given. Then a number k exists such that the inequality $|a_n - g| < \varepsilon$ holds for $n > k$. According to the formula (19), $m_n \geqslant n > k$; hence $|a_{m_n} - g| < \varepsilon$. This implies $\lim_{n=\infty} a_{m_n} = g$.

THEOREM 2 (BOLZANO-WEIERSTRASS ([1])). *Every bounded sequence contains a convergent subsequence.*

Proof. Let the sequence $\{a_n\}$ be bounded. Then a number M exists such that the inequality $-M < a_n < M$ holds for each n.

We shall now apply the theorem from § 1.7. Suppose that we denote by Z the set of numbers x for which the inequality $x < a_n$ holds for infinitely many n. The set Z is non-empty, since the number $-M$ belongs to Z (because the inequality $-M < a_n$ is satisfied by all n). At the same time this set is bounded above; for, if x belongs to Z, then $x < M$, since assuming $x \geqslant M$, the inequality $x < a_n$ would not be satisfied by any n (thus the number x could not belong to the set Z).

Since the set Z is non-empty and bounded above, the upper bound of this set exists. Let us denote this bound by g. It follows by the definition of the upper bound that to any number $\varepsilon > 0$ there exist infinitely many n such that

$$g - \varepsilon < a_n \leqslant g + \varepsilon$$

(because $g - \varepsilon$ belongs to the set Z and $g + \varepsilon$ does not belong to Z).

We shall now prove that g is the limit of a certain subsequence of the sequence $\{a_n\}$. Thus, we have to define a sequence of positive integers $m_1 < m_2 < m_3 < \ldots$ in such a way that $\lim_{n=\infty} a_{m_n} = g$.

For this purpose we shall first substitute 1 in place of ε. Then there exist infinitely many n such that

$$g - 1 < a_n \leqslant g + 1.$$

([1]) Mathematicians of the 19th century. Karl Weierstrass is one of the greatest analysts.

Denoting one of these n by m_1, we have

$$g-1 < a_{m_1} \leqslant g+1.$$

Now, let us substitute $\varepsilon = \frac{1}{2}$. Then there exist infinitely many n such that $g-\frac{1}{2} < a_n \leqslant g+\frac{1}{2}$. Among these n there exist numbers $> m_1$; someone of these n we denote by m_2. So we have

$$g-\tfrac{1}{2} < a_{m_2} \leqslant g+\tfrac{1}{2}, \quad m_1 < m_2.$$

Similarly, we can find an index m_3 such that $g-\frac{1}{3} < a_{m_3} \leqslant g+\frac{1}{3}$, $m_2 < m_3$.

In general, if m_n is already defined, we define m_{n+1} in such a way that

(21) $\quad g - \dfrac{1}{n+1} < a_{m_{n+1}} \leqslant g + \dfrac{1}{n+1}, \quad m_n < m_{n+1}.$

We shall apply formula (17) on the double inequality. Since $\lim \dfrac{1}{n} = 0$, we have

$$\lim \left(g - \frac{1}{n}\right) = g = \lim \left(g + \frac{1}{n}\right).$$

Hence the formula (21) (after replacing $n+1$ by n) implies $\lim\limits_{n=\infty} a_{m_n} = g$. Finally, it follows from the inequality $m_n < m_{n+1}$ that the sequence $a_{m_1}, a_{m_2}, \ldots$ is a subsequence of the sequence $a_1, a_2, \ldots$ Thus, the theorem is proved.

As a corollary from the above theorem we shall deduce

THEOREM 3. *Every monotone bounded sequence is convergent. Moreover, if $a_1 \leqslant a_2 \leqslant \ldots$, then $a_n \leqslant \lim a_n$; if $a_1 \geqslant a_2 \geqslant \ldots$, then $a_n \geqslant \lim a_n$.*

Let us assume the sequence $\{a_n\}$ to be bounded and increasing in the wider sense. Let us denote by Z the set of numbers belonging to this sequence and by g its upper bound (cf. § 1.7). Then we have

$$g \geqslant a_n \quad \text{for each } n;$$

at the same time for any $\varepsilon > 0$ there exists a k such that $g-\varepsilon < a_k$ (because the inequality $g-\varepsilon \geqslant a_n$ cannot hold for all n, as follows from the definition of the upper bound).

Since the sequence $\{a_n\}$ is increasing (in the wider sense), the inequality $n > k$ implies $a_k \leqslant a_n$, whence $g-\varepsilon < a_n$. Thus we conclude that to any $\varepsilon > 0$ there exists a k such that for $n > k$ the double inequality $g-\varepsilon < a_n \leqslant g$ holds and so the inequality $|a_n - g| < \varepsilon$ holds, too. This means that $g = \lim_{n=\infty} a_n$.

To prove that the inequality $a_n \leqslant g$ is satisfied by each n, let us consider a fixed positive integer n_0. It has to be proved that $a_{n_0} \leqslant g$. Now, it follows from the assumptions that

$$a_{n_0} \leqslant a_{n_0+1}, \ a_{n_0} \leqslant a_{n_0+2}, \ \ldots, \ a_{n_0} \leqslant a_{n_0+k}, \ \ldots$$

Thus by the formula (15), $a_{n_0} \leqslant \lim_{k=\infty} a_{n_0+k} = g$.

In this way the Theorem 3 is proved for non-decreasing sequences. The proof for non-increasing sequences is analogous.

It follows at once from Theorem 1 that a convergent sequence cannot contain two subsequences convergent to different limits (especially, this implies the divergence of the sequence $0, 1, 0, 1, \ldots$). For bounded sequences, this theorem may be reversed. Namely, we have

THEOREM 4 ([1]). *If a sequence $\{a_n\}$ is bounded and if all subsequences of this sequence are convergent to the same limit g, then the sequence $\{a_n\}$ is convergent to the limit g.*

In other words, every bounded divergent sequence contains two subsequences convergent to different limits.

Proof. Let us suppose that g is not the limit of the sequence $\{a_n\}$. Then a number $\varepsilon > 0$ exists such that for every k there exists $n > k$ such that the inequality $|a_n - g| \geqslant \varepsilon$ holds. Hence we conclude that a subsequence $\{a_{m_n}\}$ exists such that $|a_{m_n} - g| \geqslant \varepsilon$ for every n. By sub-

---

([1]) We shall apply this theorem in § 5.5.

stituting the value 1 in place of k we find that there exists $m_1 > 1$ such that $|a_{m_1} - g| \geqslant \varepsilon$. Further, substituting $k = m_1$ we find $m_2 > m_1$ for which $|a_{m_2} - g| \geqslant \varepsilon$. Similarly, there exists $m_3 > m_2$ such that $|a_{m_3} - g| \geqslant \varepsilon$ etc.

Since the sequence $\{a_n\}$ is bounded, the sequence $\{a_{m_n}\}$ is also bounded. Hence, by the Bolzano-Weierstrass theorem it contains a convergent subsequence. Let us denote this subsequence by $\{a_{m_{r_n}}\}$. Since it is also a subsequence of the sequence $\{a_n\}$, our assumptions imply

$$\lim_{n=\infty} a_{m_{r_n}} = g, \quad \text{i. e.} \quad \lim_{n=\infty}(a_{m_{r_n}} - g) = 0.$$

But, on the other hand, the numbers $m_{r_1}, m_{r_2}, \ldots$ belong to the sequence $m_1, m_2, \ldots$ and, consequently, the inequality $|a_{m_n} - g| \geqslant \varepsilon$ implies the inequality $|a_{m_{r_n}} - g| \geqslant \varepsilon$ for each n. This contains a contradiction.

So we have proved that $\lim_{n=\infty} a_n = g$, by leading to contradiction.

2.7. Cauchy theorem ([1])

A necessary and sufficient condition for the convergence of the sequence $\{a_n\}$ is that for every $\varepsilon > 0$ a number r exists such that the inequality

(22) $$|a_n - a_r| < \varepsilon$$

holds for $n > r$.

The proof consists of two parts. In the first one we shall prove that the convergence of a sequence implies the condition formulated in the theorem, the so-called "Cauchy condition"; in other words, we shall prove that this condition is a necessary condition for the convergence of this sequence. In the second part we shall prove that

---

[1] Augustin Cauchy, 1789-1857, one of the most eminent French mathematicians. The establishing of fundamental notions and theorems of the theory of sequences and series and of the theory of continuous functions is due mostly to him.

2. INFINITE SEQUENCES

if the Cauchy condition is satisfied, then the sequence is convergent, i.e. that it is a sufficient condition for the convergence of the sequence.

1° Let $\lim a_n = g$ and let $\varepsilon > 0$ be given. Then a number r exists such that $|a_n - g| < \frac{1}{2}\varepsilon$ holds for $n \geqslant r$. In particular, this inequality holds for $n = r$, i.e. $|a_r - g| < \frac{1}{2}\varepsilon$. Adding both these inequalities under the sign of the absolute value we obtain the formula (22).

2° Now, let us assume the Cauchy condition to be satisfied. We have to prove the convergence of the sequence. First of all we shall prove that it is a bounded sequence. The proof will be analogous to the proof in § 2.3. Namely, we substitute $\varepsilon = 1$. Then a number r exists such that we have $|a_n - a_r| < 1$ for $n > r$. Hence $|a_n| - |a_r| \leqslant |a_n - a_r| < 1$, whence $|a_n| < |a_r| + 1$. Let us denote by M a number greater than anyone among the following r numbers: $|a_1|, |a_2|, ..., |a_{r-1}|, |a_r| + 1$. So we have $M > |a_n|$ for each n. This means that the sequence $\{a_n\}$ is bounded.

Hence we conclude by the Bolzano-Weierstrass theorem that this sequence contains a convergent subsequence. Thus, let $\lim\limits_{n=\infty} a_{m_n} = g$, where $m_1 < m_2 < ...$ We shall prove that $g = \lim\limits_{n=\infty} a_n$.

Now, let $\varepsilon > 0$ be given. According to the Cauchy condition, a number r exists such that the inequality

$$(23) \qquad |a_n - a_r| < \tfrac{1}{3}\varepsilon$$

is satisfied for $n > r$.

The equation $\lim\limits_{n=\infty} a_{m_n} = g$ implies the existence of a k such that we have

$$(24) \qquad |a_{m_n} - g| < \tfrac{1}{3}\varepsilon$$

for $n > k$. Clearly, k may be chosen in such way that $k > r$. Then both inequalities (23) and (24) are satisfied for $n > k$.

Moreover, one may replace in formula (23) n by m_n, since $m_n \geqslant n > r$ (cf. (19)). This gives

(25) $\qquad |a_{m_n} - a_r| < \tfrac{1}{3}\varepsilon$.

Adding the inequalities (23)-(25) after changing the sign under the absolute value in (25), we obtain the inequality $|a_n - g| < \varepsilon$, valid for each $n > k$. This means that $\lim a_n = g$.

Remarks. (α) In the Cauchy theorem as well as in the definition of the limit the sign $>$ in the inequality $n > k$ as well as the sign $<$ in the formula (22) may be replaced by $\geqslant$ and $\leqslant$, respectively.

(β) The Cauchy condition may be also formulated as follows: *for every $\varepsilon > 0$ a number r exists such that the conditions $n > r$ and $n' > r$ imply $|a_n - a_{n'}| < \varepsilon$.*

Namely, the Cauchy condition implies the existence of an r such that the inequalities $|a_n - a_r| < \tfrac{1}{2}\varepsilon$ and $|a_{n'} - a_r| < \tfrac{1}{2}\varepsilon$ hold for $n > r$ and $n' > r$. Adding these inequalities we obtain $|a_n - a_{n'}| < \varepsilon$.

2.8. Divergence to ∞

A sequence $\{a_n\}$ is called *divergent to* ∞, if to every number r there exists a positive integer k such that $a_n > r$ for $n > k$. We denote this fact symbolically by the equation: $\lim\limits_{n=\infty} a_n = \infty$. Thus, a sequence divergent to ∞ is a sequence, in which those terms with sufficiently large indices are arbitrarily great.

Analogously we define the meaning of the expression $\lim\limits_{n=\infty} a_n = -\infty$. It means that for every negative number r there exists a k such that $a_n < r$ for $n > k$.

We say that sequences divergent to ∞ or $-\infty$ have *improper* limits.

EXAMPLES: $\lim\limits_{n=\infty} n = \infty$, $\lim\limits_{n=\infty} (-n) = -\infty$.

1. *An increasing (in the wider sense) sequence, unbounded above, is divergent to* ∞.

Indeed, since the sequence $\{a_n\}$ is unbounded above, there exists for any r a number k such that $a_n \geqslant a_k$ for $n > k$, whence $a_n > r$. Thus, $\lim\limits_{n=\infty} a_n = \infty$.

An analogous theorem holds for decreasing sequences.

Comparing this theorem with Theorem 3, § 2.6, we conclude that *every monotone sequence has a limit, proper or improper* (depending on whether it is bounded or unbounded, respectively).

2. *If* $\lim\limits_{n=\infty} a_n = \pm\infty$, *then* $\lim\limits_{n=\infty} (1/a_n) = 0$.

Namely, let $\lim a_n = \infty$ and let $\varepsilon > 0$. Writing $r = 1/\varepsilon$ we conclude that there exists a k such that $a_n > 1/\varepsilon$ holds for $n > k$, i.e. $1/a_n < \varepsilon$ for $n > k$; but this means that $\lim (1/a_n) = 0$.

The converse theorem is not true: the sequence $1, -\tfrac{1}{2}, \tfrac{1}{3}, -\tfrac{1}{4}, \ldots$ converges to 0 but the sequence of its reciprocals $1, -2, 3, -4, \ldots$ is divergent neither to $+\infty$ nor to $-\infty$.

Let us mention the following theorem concerning operations on divergent sequences: *the sum and the product of two sequences divergent to $+\infty$ are divergent to $+\infty$.* We shall prove this theorem in the following form which is a little stronger than the above:

3. *If* $\lim a_n = \infty$ *and if the sequence $\{b_n\}$ is bounded below, then* $\lim (a_n + b_n) = \infty$.

Namely, let $M < b_n$. Since $\lim a_n = \infty$, then for a given r there exists a number k such that we have $a_n > r - M$ for $n > k$. Hence $a_n + b_n > r$ and, consequently, $\lim (a_n + b_n) = \infty$.

4. *If* $\lim a_n = \infty$ *and if* $b_n \geqslant c$ *always, where* $c > 0$, *then* $\lim (a_n \cdot b_n) = \infty$.

Let a number $r > 0$ be given. There exists a number k such that for $n > k$, we have $a_n > r/c$. Multiplying this inequality by the inequality $b_n \geqslant c$ we obtain $a_n b_n > r$. Hence $\lim (a_n \cdot b_n) = \infty$.

A theorem corresponding to theorem (15) is the following one:

5. *If* $\lim a_n = \infty$ *and* $a_n \leqslant b_n$, *then* $\lim b_n = \infty$.

Namely, if $a_n > r$, then still more $b_n > r$.

2.9. Examples

(α) *If* $c > 0$, *then* $\lim\limits_{n=\infty} nc = \infty$.

This follows immediately from 4.

(β) *If* $a > 1$, *then* $\lim\limits_{n=\infty} a^n = \infty$.

Namely, writing $c = a-1$, we get $c > 0$, whence $\lim\limits_{n=\infty} nc = \infty$. On the other hand, applying the Bernoulli inequality (§ 1 (1)), we obtain

$$a^n = (1+c)^n \geqslant 1 + nc, \quad \text{whence} \quad \lim\limits_{n=\infty} a^n = \infty$$

according to 5.

(γ) *If* $|q| < 1$, *then* $\lim q^n = 0$.

First, let us assume that $0 < q < 1$. Then $1/q > 1$ and, consequently, $\lim\limits_{n=\infty}(1/q^n) = \infty$. Hence we obtain, according to Theorem 2, $\lim\limits_{n=\infty} q^n = 0$.

Hence it follows that in the general case, assuming only $|q| < 1$, the equation $\lim\limits_{n=\infty}|q|^n = 0$ is satisfied, i. e. $\lim\limits_{n=\infty}|q^n| = 0$. But the last equation implies $\lim\limits_{n=\infty} q^n = 0$ (cf. (18)).

(δ) *If* $a > 0$, *then* $\lim\limits_{n=\infty} \sqrt[n]{a} = 1$.

First, let us assume that $a > 1$ and let us write $u_n = \sqrt[n]{a}$. We shall prove that the sequence $\{u_n\}$ is decreasing. Indeed, $a > 1$ implies $a^n < a^{n+1}$. Taking on both sides the $n(n+1)$-th root we obtain $\sqrt[n+1]{a} < \sqrt[n]{a}$, i. e. $u_n > u_{n+1}$. Moreover, the sequence $\{u_n\}$ is bounded since $u_n > 1$. Hence we conclude that it is a convergent sequence (cf. 2.6, Theorem 3). One has to prove that 1 is the limit of this sequence. Let us write $\lim u_n = 1+h$. By the above cited theorem, $u_n > 1$ implies $h \geq 0$. We shall prove that the supposition $h > 0$ leads to contradiction.

Indeed, the inequality $1+h < \sqrt[n]{a}$ yields $(1+h)^n < a$ and since $(1+h)^n > 1+nh$, we conclude that $1+nh < a$ for each n. But this is impossible, since $\lim_{n=\infty} nh = \infty$.

So we have proved that $a > 1$ implies $\lim_{n=\infty} \sqrt[n]{a} = 1$. But if $0 < a \leq 1$, then $1/a \geq 1$ and $\lim_{n=\infty} \sqrt[n]{1/a} = 1$, i.e. $\lim_{n=\infty} (1/\sqrt[n]{a}) = 1$. Hence $1/\lim_{n=\infty}\sqrt[n]{a} = 1$, i.e. $\lim_{n=\infty}\sqrt[n]{a} = 1$.

(ε) *If* $|q| < 1$, *then* $\lim_{n=\infty} a(1+q+q^2+\ldots+q^{n-1}) = \dfrac{a}{1-q}$.

This is the expression for the sum of the geometric progression with a quotient whose numerical value is less than unity.

Namely, we have $1+q+q^2+\ldots+q^{n-1} = \dfrac{1-q^n}{1-q}$ and $\lim q^n = 0$ (cf. (γ)).

(ζ) *If* $a > 0$ *and if* $\{r_n\}$ *is a sequence of rational numbers convergent to zero, then* $\lim_{n=\infty} a^{r_n} = 1$.

This is a generalization of the formula (δ) and it is easily deduced from this formula.

2.10. The number e

We shall prove the existence of $\lim_{n=\infty}\left(1+\dfrac{1}{n}\right)^n$. We shall denote this limit by e.

Let us write $a_n = \left(1+\dfrac{1}{n}\right)^n$. Then we have

$a_1 = 2$, $\quad a_2 = (1+\tfrac{1}{2})^2 = \tfrac{9}{4}$, $\quad a_3 = (1+\tfrac{1}{3})^3 = \tfrac{64}{27} = 2\tfrac{10}{27}$.

We shall prove that this sequence is increasing. For this purpose we shall apply the Newton binomial formula (§ 1.3):

$$a_n = 1 + n\frac{1}{n} + \frac{n(n-1)}{1\cdot 2}\cdot\frac{1}{n^2} + \ldots + \frac{n(n-1)\cdot\ldots\cdot 1}{1\cdot 2\cdot\ldots\cdot n}\cdot\frac{1}{n^n}$$

$$= 1 + 1 + \frac{1-\frac{1}{n}}{2!} + \frac{\left(1-\frac{1}{n}\right)\left(1-\frac{2}{n}\right)}{3!} + \ldots + $$

$$+ \frac{\left(1-\frac{1}{n}\right)\cdot\ldots\cdot\left(1-\frac{n-1}{n}\right)}{n!},$$

hence

$$a_{n+1} = 1 + 1 + \frac{1-\frac{1}{n+1}}{2!} + \frac{\left(1-\frac{1}{n+1}\right)\left(1-\frac{2}{n+1}\right)}{3!} + \ldots +$$

$$+ \frac{\left(1-\frac{1}{n+1}\right)\cdot\ldots\cdot\left(1-\frac{n-1}{n+1}\right)}{n!} + \frac{\left(1-\frac{1}{n+1}\right)\cdot\ldots\cdot\left(1-\frac{n}{n+1}\right)}{(n+1)!}.$$

It is easily seen that the terms of the sum constituting a_n are less than or equal to the first terms constituting a_{n+1}, respectively. Hence $a_n < a_{n+1}$.

Thus, the sequence is increasing. But it is also bounded, for

$$a_n < 2 + \frac{1}{2!} + \ldots + \frac{1}{n!} < 2 + \frac{1}{2} + \frac{1}{2^2} + \ldots + \frac{1}{2^{n-1}} < 3,$$

since (cf. § 2.9, (ε))

$$\frac{1}{2} + \frac{1}{2^2} + \ldots + \frac{1}{2^{n-1}} < \frac{\frac{1}{2}}{1-\frac{1}{2}} = 1.$$

The sequence $\{a_n\}$, being increasing and bounded, is convergent. Hence the existence of the number e is already proved. It can be calculated that, approximately, $e = 2{,}718\ldots$ We may also prove (as we shall show later) that

$$e = \lim_{n\to\infty}\left(1 + \frac{1}{1!} + \frac{1}{2!} + \ldots + \frac{1}{n!}\right).$$

The number e plays a great part in analysis. In particular it is used as the base of logarithms (so called *natural logarithms*).

We shall now prove that

(26) $$\lim_{n=\infty}\left(1-\frac{1}{n}\right)^n = \frac{1}{e}.$$

Indeed, we have

$$1-\frac{1}{n} = \frac{n-1}{n} = \frac{1}{\left(\dfrac{n}{n-1}\right)} = \frac{1}{1+\dfrac{1}{n-1}},$$

whence

$$\lim_{n=\infty}\left(1-\frac{1}{n}\right)^n = \frac{1}{\lim\limits_{n=\infty}\left(1+\dfrac{1}{n-1}\right)^n}$$

$$= \frac{1}{\lim\limits_{n=\infty}\left(1+\dfrac{1}{n-1}\right)^{n-1}} \cdot \frac{1}{\lim\limits_{n=\infty}\left(1+\dfrac{1}{n-1}\right)} = \frac{1}{e},$$

for

$$\lim_{n=\infty}\left(1+\frac{1}{n-1}\right) = 1.$$

2.11*. The sequences of the arithmetic means and of the geometric means of a given sequence

Given a sequence $a_1, a_2, \ldots, a_n, \ldots$, let us consider the sequence of the arithmetic means

$$a_1, \frac{a_1+a_2}{2}, \ldots, \frac{a_1+a_2+\ldots+a_n}{n}, \ldots$$

1. *If* $\lim\limits_{n=\infty} a_n = g$, *then* $\lim\limits_{n=\infty}\dfrac{a_1+\ldots+a_n}{n} = g$; *this theorem remains true in the case* $g = \pm\infty$.

Proof. First, let us assume that g is a finite number. Then, for a given $\varepsilon > 0$, a number k exists such that the inequalities

$$g-\varepsilon < a_{k+1} < g+\varepsilon, \quad g-\varepsilon < a_{k+2} < g+\varepsilon,$$
$$\ldots, g-\varepsilon < a_n < g+\varepsilon$$

are satisfied for each $n > k$.

Adding the system of these $n-k$ inequalities we obtain

$$g-\varepsilon < \frac{a_{k+1}+\ldots+a_n}{n-k} < g+\varepsilon,$$

i.e.

$$(g-\varepsilon)\frac{n-k}{n} < \frac{a_{k+1}+\ldots+a_n}{n} < (g+\varepsilon)\frac{n-k}{n}.$$

At the same time, $\lim_{n=\infty} \frac{a_1+\ldots+a_k}{n} = 0$, whence we have $-\varepsilon < \frac{a_1+\ldots+a_k}{n} < \varepsilon$ for sufficiently large n. Hence

$$(g-\varepsilon)\frac{n-k}{n} - \varepsilon < \frac{a_1+\ldots+a_n}{n} < (g+\varepsilon)\frac{n-k}{n} + \varepsilon,$$

but since $\lim_{n=\infty} \frac{n-k}{n} = 1$, we have

$$g-3\varepsilon < \frac{a_1+\ldots+a_n}{n} < g+3\varepsilon$$

for sufficiently large n.

This means that $\lim_{n=\infty} \frac{a_1+\ldots+a_n}{n} = g$.

Now, let us assume that $g = \infty$ (the proof in the case $g = -\infty$ is analogous). Then for a given number $r > 0$ a number k exists such that $a_{k+1} > r$, $a_{k+2} > r$, ... Thus, we have $a_{k+1}+\ldots+a_n > (n-k)r$ for each $n > k$, whence

$$\frac{a_{k+1}+\ldots+a_n}{n} > \frac{n-k}{n}r.$$

But we have $\frac{a_1+\ldots+a_k}{n} > -\frac{r}{2}$ for large n; whence

$$\frac{a_1+\ldots+a_n}{n} > \frac{n-k}{n}r - \frac{r}{2} = \frac{r}{2} - \frac{k}{n}r,$$

whence we obtain $\dfrac{a_1+\ldots+a_n}{n} > \dfrac{r}{3}$ for sufficiently large n. Hence we conclude that

$$\lim_{n=\infty} \frac{a_1+\ldots+a_n}{n} = \infty.$$

Remark. The converse theorem is not true: a sequence may be divergent although the sequence of its arithmetic means is convergent. For instance, the sequence $1, 0, 1, 0, \ldots$ is divergent but the sequence of its arithmetic means is convergent to $\tfrac{1}{2}$.

An analogous theorem holds for the geometric means:

2. *If $a_n > 0$ and $\lim\limits_{n=\infty} a_n = g$ (g finite or infinite), then*

$$\lim_{n=\infty} \sqrt[n]{a_1 a_2 \cdot \ldots \cdot a_n} = g.$$

This theorem is proved in a manner completely analogous to the proof of Theorem 1. Theorem 2 may be also reduced to Theorem 1 by taking logarithms (applying the continuity of the function $\log x$, cf. § 5.5).

3. *If $\lim\limits_{n=\infty}(a_{n+1}-a_n) = g$, then $\lim\limits_{n=\infty}\dfrac{a_n}{n} = g$.*

Indeed, let us write $b_n = a_n - a_{n-1}$ for $n > 1$ and $b_1 = a_1$. Then we have

$$\lim_{n=\infty} b_n = g \quad \text{and} \quad \frac{b_1+\ldots+b_n}{n} = \frac{a_n}{n}.$$

Hence $\lim\limits_{n=\infty} \dfrac{a_n}{n} = g$ by Theorem 1.

4. *If $a_n > 0$ and $\lim\limits_{n=\infty}\dfrac{a_{n+1}}{a_n} = g$, then $\lim\limits_{n=\infty}\sqrt[n]{a_n} = g$.*

This theorem is obtained from Theorem 2 writing $b_n = \dfrac{a_n}{a_{n-1}}$ for $n > 1$ and $b_1 = a_1$; it follows from the equation $\sqrt[n]{b_1 b_2 \cdot \ldots \cdot b_n} = \sqrt[n]{a_n}$.

EXAMPLES. (α) Writing $a_n = n$, we have $\lim\limits_{n=\infty} \dfrac{n+1}{n} = 1$. Hence $\lim\limits_{n=\infty} \sqrt[n]{n} = 1$.

(β) $\lim\limits_{n=\infty} \sqrt[n]{n!} = \infty$, for $\lim\limits_{n=\infty} \dfrac{(n+1)!}{n!} = \lim\limits_{n=\infty} (n+1) = \infty$.

(γ) The example of the sequence $1, 1, \tfrac{1}{2}, \tfrac{1}{2}, \tfrac{1}{4}, \tfrac{1}{4}, \ldots$ shows that the limit $\lim\limits_{n=\infty} \sqrt[n]{a_n}$ may exist, although the limit $\lim\limits_{n=\infty} \dfrac{a_{n+1}}{a_n}$ does not exist.

Exercises on § 2

1. Evaluate the limits:

$$\lim_{n=\infty} \frac{n}{n+1}, \quad \lim_{n=\infty} \frac{5n+1}{7n-2},$$

$$\lim_{n=\infty} \frac{2n^2+5}{3n^3-1}, \quad \lim_{n=\infty} \frac{1^2+2^2+\ldots+n^2}{n^3}.$$

2. Let a convergent sequence $\{a_n\}$ be given. Prove that, if we replace a finite number of terms of this sequence by other terms, the sequence obtained in this way is convergent, too, and has the same limit as the given sequence.

Similarly: cancellation or addition of a finite number of terms to a sequence has influence neither on the convergence of the sequence nor on the value of its limit.

3. Prove that if $\lim\limits_{n=\infty} a_n = 0$ and if the sequence $\{b_n\}$ is bounded, then $\lim\limits_{n=\infty} a_n b_n = 0$.

4. Prove that if the sequences $\{a_n\}$ and $\{b_n\}$ are convergent to the same limit, then the sequence $a_1, b_1, a_2, b_2, \ldots$ is convergent to this limit, too.

5. Prove the following generalization of the theorem on the limit of subsequences (§ 2.6, Theorem 1): *if* $\lim\limits_{n=\infty} a_n = g$

(where g has a finite or infinite value) and if $\lim_{n=\infty} m_n = \infty$, *then* $\lim_{n=\infty} a_{m_n} = g$.

6. Evaluate $\lim_{n=\infty} \frac{1}{n} \sqrt[n]{n!}$.

7. We define a sequence $\{a_n\}$ by induction as follows:
$$a_1 = 0, \quad a_2 = 1, \quad a_n = \frac{a_{n-1} + a_{n-2}}{2}.$$
Prove that $\lim_{n=\infty} a_n = \frac{2}{3}$.

8. Let a bounded sequence $\{a_n\}$ be given. Prove that among the numbers being limits of convergent subsequences of this sequence there exist the greatest number and the least number (these numbers are called *limes superior* and *limes inferior* of the sequence $\{a_n\}$, i.e. the upper limit and the lower limit of the sequence $\{a_n\}$, respectively).

Prove that a necessary and sufficient condition of the convergence of a sequence is that the upper limit equals the lower limit.

Prove that if $c = \limsup a_n$, then for every $\varepsilon > 0$ a number k exists such that $a_n < c + \varepsilon$ for $n > k$ and that $a_m > c - \varepsilon$ holds for infinitely many indices m.

9. Prove that for every sequence of closed intervals such that each interval is contained in the preceding one, there exists at least one common point (Ascoli theorem).

In the above theorem closed intervals are considered, i.e. intervals together with end-points; show by an example that this theorem does not hold for open intervals (i.e. intervals whose end-points are excluded).

10. Find the limit of the sequence $\{a_n\}$ defined by induction:
$$a_1 = \sqrt{c}, \quad a_{n+1} = \sqrt{c + a_n}.$$

11. Prove that if $\lim\limits_{n=\infty} a_n = g$ and if $\lim\limits_{n=\infty} (b_1 + b_2 + \ldots + b_n) = \infty$, where always $b_n \geqslant 0$, then

$$\lim_{n=\infty} \frac{a_1 b_1 + a_2 b_2 + \ldots + a_n b_n}{b_1 + b_2 + \ldots + b_n} = g.$$

12. Prove that if $\lim\limits_{n=\infty} \dfrac{u_n - u_{n-1}}{v_n - v_{n-1}} = g$ and $\lim\limits_{n=\infty} v_n = \infty$, where the sequence $\{v_n\}$ is increasing, then $\lim\limits_{n=\infty} \dfrac{u_n}{v_n} = g$.

(Substitute in the previous exercise: $a_n = \dfrac{u_n - u_{n-1}}{v_n - v_{n-1}}$, $b_n = v_n - v_{n-1}$).

13. Pascal ([1]) triangle. We consider the following table (i.e. double sequence $\{a_{nk}\}$):

$$\begin{array}{ccccc} 1 & 1 & 1 & 1 & 1 \ldots \\ 1 & 2 & 3 & 4 & 5 \ldots \\ 1 & 3 & 6 & 10 & 15 \ldots \\ 1 & 4 & 10 & 20 & 35 \ldots \\ 1 & 5 & 15 & 35 & 70 \ldots \\ \multicolumn{5}{c}{\ldots\ldots\ldots\ldots\ldots} \end{array}$$

This table is defined as follows: the first line consists of nothing but 1, i.e. $a_{1,k} = 1$; the k-th term in the n-th line is the sum of the first k terms of the $(n-1)$-th line, i. e.

$$a_{nk} = a_{n-1,1} + a_{n-1,2} + \ldots + a_{n-1,k}.$$

Prove that

$$a_{nk} = \frac{(k+n-2)!}{(k-1)!(n-1)!}$$

and that the terms $a_{n,1}, a_{n-1,2}, \ldots, a_{1,n}$ are successive coefficients of the Newton expansion of the expression $(a+b)^{n-1}$ (as is easily seen they are the terms of the table lying on the straight line joining the n-th term of the first line with the first term of the n-th line).

([1]) Blaise Pascal, a French mathematician, 1623-1662, known by his works which initiated the Calculus of Probability. Newton's predecessor in certain researches in Analysis.

14. Prove that
$$\lim_{n=\infty} \left(\frac{1}{\sqrt{n^2+1}} + \frac{1}{\sqrt{n^2+2}} + \ldots + \frac{1}{\sqrt{n^2+n}} \right) = 1.$$

15. Prove that if a sequence of rational numbers $\left\{\frac{p_n}{q_n}\right\}$ (where p_n and q_n are positive integers) tends to an irrational number, then
$$\lim_{n=\infty} p_n = \infty = \lim_{n=\infty} q_n.$$

§ 3. INFINITE SERIES

3.1. Definitions and examples

Given an infinite sequence $a_1, a_2, \ldots, a_n, \ldots$, we consider the following infinite sequence:

(1) $s_1 = a_1, \; s_2 = a_1 + a_2, \; \ldots, \; s_n = a_1 + a_2 + \ldots + a_n, \; \ldots$

If the sequence $s_1, s_2, \ldots, s_n, \ldots$ has a limit, then we denote this limit by the symbol
$$\sum_{n=1}^{\infty} a_n = \lim_{n=\infty} s_n$$
and we call it the *sum of the infinite series* $a_1 + a_2 + \ldots + a_n + \ldots$

In this case we say also that the *series is convergent*. If the above limit does not exist, then the series is called *divergent*.

In the case when all terms of the sequence $\{a_n\}$ beginning with a certain k are $= 0$ we have an ordinary sum of a finite number of terms. Thus, the sum of an infinite series is a generalization of this sum when the number of terms is infinite (which is emphasized by the notation). However, in distinction with the addition of a finite number of terms, it is not always possible to find the sum of an infinite number of terms, i.e. the sequence (1) is not always convergent (even if we take into account infinite limits).

To illustrate this we shall give several examples.

The geometric series $a + aq + aq^2 + \ldots + aq^n + \ldots$ with the quotient q satisfying the inequality $|q| < 1$ is convergent. Namely (§ 2.9, (ε)),

$$a + aq + aq^2 + \ldots = \frac{a}{1-q}.$$

The series $1 + 1 + 1 + \ldots$ is divergent to ∞ because $s_n = n$.

The series $1 - 1 + 1 - 1 + \ldots$ has neither a proper nor an improper limit, for in this case the sequence (1) is the sequence: $1, 0, 1, 0, \ldots$

The sequence $\{s_n\}$ will be called the *sequence of partial sums* of the given infinite series.

By the *n-th remainder* of the series $a_1 + a_2 + \ldots$ we understand the series

(2) $$r_n = a_{n+1} + a_{n+2} + \ldots = \sum_{m=n+1}^{\infty} a_m.$$

If the series $a_1 + a_2 + \ldots$ is convergent, then $\lim_{n=\infty} r_n = 0$.

First of all let us note that if the series $a_1 + a_2 + \ldots$ is convergent, then the series (2) is also convergent for each value of n. For

$$r_n = \lim_{k=\infty}(a_{n+1} + a_{n+2} + \ldots + a_{n+k}) = \lim_{k=\infty} s_{n+k} - s_n = \sum_{m=1}^{\infty} a_m - s_n,$$

whence

$$\lim_{n=\infty} r_n = \sum_{m=1}^{\infty} a_m - \lim_{n=\infty} s_n = \sum_{m=1}^{\infty} a_m - \sum_{m=1}^{\infty} a_m = 0.$$

3.2. General properties of series

We shall now give some properties of infinite series which are direct consequences of relevant properties of infinite sequences.

3. INFINITE SERIES

1. CAUCHY CONDITION. *A necessary and sufficient condition for the convergence of the series $a_1 + a_2 + \ldots$ is that for every $\varepsilon > 0$ a number k exists such that the inequality*

(3) $$|a_{k+1} + a_{k+2} + \ldots + a_{k+m}| < \varepsilon$$

holds for each m.

Indeed, the convergence of the given series means the convergence of its partial sums $s_1, s_2, \ldots$, and the Cauchy theorem (§ 2.7) implies this sequence to be convergent if and only if for every $\varepsilon > 0$ there is a k such that for each m the inequality $|s_{k+m} - s_k| < \varepsilon$ (i.e. the inequality (3)) is satisfied.

2. *If the series $a_1 + a_2 + \ldots$ is convergent, then $\lim_{n=\infty} a_n = 0$.*

Namely, $a_n = s_n - s_{n-1}$; hence

$$\lim_{n=\infty} a_n = \lim_{n=\infty} s_n - \lim_{n=\infty} s_{n-1},$$

but

$$\lim_{n=\infty} s_n = \lim_{n=\infty} s_{n-1}.$$

Remark. Theorem 2 cannot be reversed. Namely, *there exist divergent series with terms tending to 0*. Such a series is the harmonic series

$$1 + \frac{1}{2} + \frac{1}{3} + \ldots + \frac{1}{n} + \ldots$$

Indeed,

$$\frac{1}{3} + \frac{1}{4} > \frac{1}{2}, \quad \frac{1}{5} + \frac{1}{6} + \frac{1}{7} + \frac{1}{8} > \frac{1}{2}, \ldots,$$

$$\frac{1}{2^n+1} + \frac{1}{2^n+2} + \ldots + \frac{1}{2^{n+1}} > \frac{1}{2},$$

i.e.

$$s_{2^{n+1}} - s_{2^n} > \tfrac{1}{2}.$$

Hence $s_{2^n} > \dfrac{n}{2}$. Thus $\lim_{n=\infty} s_{2^n} = \infty$, whence we conclude that the sequence of partial sums of the harmonic series and so the series itself, are divergent. This is a series divergent to ∞.

A series is called *bounded* if the sequence of its partial sums is bounded, i.e. if a number M exists such that $M > |s_n|$ for each n.

3. Every convergent series is bounded.

This is an immediate consequence of theorem § 2.3. We shall give now two theorems relating to operations on series:

4. If the series $a_1 + a_2 + \ldots$ and $b_1 + b_2 + \ldots$ are convergent, then

$$\sum_{n=1}^{\infty}(a_n + b_n) = \sum_{n=1}^{\infty} a_n + \sum_{n=1}^{\infty} b_n,$$

$$\sum_{n=1}^{\infty}(a_n - b_n) = \sum_{n=1}^{\infty} a_n - \sum_{n=1}^{\infty} b_n.$$

For

$$\sum_{n=1}^{\infty}(a_n + b_n) = \lim_{n=\infty}(a_1 + b_1 + a_2 + b_2 + \ldots + a_n + b_n)$$

$$= \lim_{n=\infty}(a_1 + a_2 + \ldots + a_n) + \lim_{n=\infty}(b_1 + b_2 + \ldots + b_n)$$

$$= \sum_{n=1}^{\infty} a_n + \sum_{n=1}^{\infty} b_n.$$

The proof for the difference is analogous.

5. If the series $a_1 + a_2 + \ldots$ is convergent, then

$$\sum_{n=1}^{\infty}(c \cdot a_n) = c \cdot \sum_{n=1}^{\infty} a_n.$$

In particular,

$$\sum_{n=1}^{\infty}(-a_n) = -\sum_{n=1}^{\infty} a_n.$$

For

$$\sum_{n=1}^{\infty}(c \cdot a_n) = \lim_{n=\infty}(c \cdot a_1 + c \cdot a_2 + \ldots + c \cdot a_n)$$

$$= c \cdot \lim_{n=\infty}(a_1 + a_2 + \ldots + a_n) = c \cdot \sum_{n=1}^{\infty} a_n.$$

3.3. Alternating series. Abel theorem

We call *alternating series* a series of the form
(4) $\quad a_1 - a_2 + a_3 - a_4 + \ldots, \quad$ where $a_n \geqslant 0$.

THEOREM 1. *An alternating series* (1) *satisfying the conditions*
(5) $\quad a_1 \geqslant a_2 \geqslant a_3 \geqslant \ldots \quad$ and $\quad \lim\limits_{n=\infty} a_n = 0$

is convergent. Moreover, the partial sums $s_n = a_1 - a_2 + a_3 - a_4 + \ldots \pm a_n$ *satisfy the inequalities*
(6) $\quad s_{2n} \leqslant a_1 - a_2 + a_3 - \ldots \leqslant s_{2n+1}$.

Indeed, the sequence of partial sums with even indices is increasing (in the wider sense), for

$$s_{2n+2} = s_{2n} + (a_{2n+1} - a_{2n+2}) \quad \text{and} \quad a_{2n+1} - a_{2n+2} \geqslant 0.$$

At the same time, it is a bounded sequence, since

$$s_{2n} = a_1 - [(a_2 - a_3) + (a_4 - a_5) + \ldots + a_{2n}] \leqslant a_1.$$

Thus, it is a convergent sequence. Let $\lim\limits_{n=\infty} s_{2n} = g$. To prove the convergence of the series (4), it suffices to show that $\lim\limits_{n=\infty} s_{2n+1} = g$.

Now, $s_{2n+1} = s_{2n} + a_{2n+1}$, whence $\lim\limits_{n=\infty} s_{2n+1} = \lim\limits_{n=\infty} s_{2n} + \lim\limits_{n=\infty} a_{2n+1} = g$, because $\lim\limits_{n=\infty} a_{2n+1} = 0$ by (5).

Finally, the double inequality (6) follows from the fact that the sequence $\{s_{2n}\}$ is increasing and the sequence $\{s_{2n+1}\}$ is decreasing (cf. § 2.6, Theorem 3).

EXAMPLES. The (anharmonic) series
(7) $\quad 1 - \tfrac{1}{2} + \tfrac{1}{3} - \tfrac{1}{4} + \ldots$

is convergent. As we shall show later, its sum equals to $\log_e 2$.

The series
(8) $\quad 1 - \tfrac{1}{3} + \tfrac{1}{5} - \tfrac{1}{7} + \ldots$

is also convergent (its sum is equal to $\tfrac{1}{4}\pi$).

THEOREM 2 (ABEL ([1])). *If the sequence $\{a_n\}$ satisfies the conditions* (5) *and if the series $b_1 + b_2 + ...$ (convergent or divergent) is bounded, then the series*

(9) $$a_1 b_1 + a_2 b_2 + ... + a_n b_n + ...$$

is convergent.

Let us write $s_n = b_1 + b_2 + ... + b_n$. According to the assumptions we have made there exists an M such that $|s_n| < M$.

To prove the convergence of the series (9), we have to estimate the sum

$$a_k b_k + a_{k+1} b_{k+1} + ... + a_n b_n$$

for $n > k$. For this purpose we note first that

(10) $$-2M < b_m + b_{m+1} + ... + b_n < 2M$$

for each m and $n \geqslant m$, since

$$|b_m + ... + b_n| = |s_n - s_{m-1}| \leqslant |s_n| + |s_{m-1}| < 2M.$$

Now,

(11) $a_k b_k + a_{k+1} b_{k+1} + ... + a_n b_n$
$= (a_k - a_{k+1}) b_k + (a_{k+1} - a_{k+2})(b_k + b_{k+1}) +$
$+ (a_{k+2} - a_{k+3})(b_k + b_{k+1} + b_{k+2}) + ... + a_n(b_k + b_{k+1} +$
$+ ... + b_n) \leqslant 2M[(a_k - a_{k+1}) + (a_{k+1} - a_{k+2}) + ... + a_n]$
$= 2M a_k$

according to (10).

Similarly, $a_k b_k + ... + a_n b_n \geqslant -2M \cdot a_k$; hence

(12) $$|a_k b_k + ... + a_n b_n| \leqslant 2M \cdot a_k.$$

Now, let an $\varepsilon > 0$ be given. By the equation (5), there exists a k such that $a_k < \dfrac{\varepsilon}{2M}$. As we have proved, the inequality (12), i. e. $|a_k b_k + ... + a_n b_n| < \varepsilon$, holds for each $n > k$. According to the Cauchy theorem (§ 3.2, Theorem 1), this implies the convergence of the series $\sum\limits_{n=1}^{\infty} a_n b_n$.

([1]) Niels Henrik Abel (1802-1829), a famous Norwegian algebraician.

Remarks. (α) It follows from the formula (12) that

(13) $$\left|\sum_{n=1}^{\infty} a_n b_n\right| \leqslant 2M \cdot a_1.$$

Namely, by the formula (12), the inequality $|a_1 b_1 + ... + a_n b_n| \leqslant 2M \cdot a_1$ holds for each n.

(β) Theorem 1 is a special case of Theorem 2. Namely, we take the sequence $1, -1, 1, -1, ...$ in place of the sequence $\{b_n\}$.

(γ) In the formula (11) we have applied the so called *Abel transformation*. It may be written also in the following form:

$$u_1 v_1 + ... + u_m v_m = u_1(v_1 - v_2) + (u_1 + u_2)(v_2 - v_3) + ... + (u_1 + ... + u_{m-1})(v_{m-1} - v_m) + (u_1 + ... + u_m) v_m.$$

3.4. Series with positive terms. D'Alembert and Cauchy convergence criterions

Assuming all the terms of the series $a_1 + a_2 + ...$ to be positive, the sequence of its partial sums is increasing. Hence it follows immediately that:

1. *A series with positive terms is either convergent or divergent to* ∞.

2. THEOREM ON COMPARISON OF SERIES. *If* $0 \leqslant b_n \leqslant a_n$ *always and if the series* $a_1 + a_2 + ...$ *is convergent, then the series* $b_1 + b_2 + ...$ *is also convergent. Moreover,*

$$\sum_{n=1}^{\infty} b_n \leqslant \sum_{n=1}^{\infty} a_n.$$

Let $s_n = a_1 + a_2 + ... + a_n$, $t_n = b_1 + b_2 + ... + b_n$. Then we have $t_n \leqslant s_n$ and, since (cf. § 2.6, 3) $s_n \leqslant \lim_{n=\infty} s_n = \sum_{n=1}^{\infty} a_n$, we have $t_n \leqslant \sum_{n=1}^{\infty} a_n$. This inequality proves the boundedness and so the convergence of the series $b_1+b_2+...$ (by Theorem 1); on the other hand, it implies the inequality $\sum_{n=1}^{\infty} b_n \leq \sum_{n=1}^{\infty} a_n$ (by § 2.6, 3).

Theorem 2 often makes it possible to prove the convergence of a series in a simple way, by comparison of this series with a series the convergence of which is already established. For example to prove the convergence of the series

$$(14) \qquad \frac{1}{2^2}+\frac{1}{3^2}+...+\frac{1}{(n+1)^2}+...\,^{(1)},$$

we compare it with the series

$$(15) \qquad \frac{1}{1\cdot 2}+\frac{1}{2\cdot 3}+...+\frac{1}{n(n+1)}+...$$

Namely, we have

$$\frac{1}{2^2} \leq \frac{1}{1\cdot 2},\ \frac{1}{3^2} \leq \frac{1}{2\cdot 3},\ ...,\ \frac{1}{(n+1)^2} \leq \frac{1}{n(n+1)}.$$

At the same time the convergence of the series (15) follows immediately from the identity:

$$\frac{1}{1\cdot 2}+\frac{1}{2\cdot 3}+...+\frac{1}{(n-1)n} = \left(1-\frac{1}{2}\right)+\left(\frac{1}{2}-\frac{1}{3}\right)+...+$$
$$+\left(\frac{1}{n-1}-\frac{1}{n}\right) = 1-\frac{1}{n},$$

whence $\lim_{n=\infty} s_n = 1$.

The following two convergence criteria for series with positive terms are based on the comparison of the given series with the geometric series.

---

[1] We shall show later (§ 11, (46)) that $1+\dfrac{1}{2^2}+\dfrac{1}{3^2}+... = \dfrac{\pi^2}{6}$.

3. D'ALEMBERT ([1]) CRITERION. *A series $a_1 + a_2 + ...$ with positive terms satisfying the condition*

(16) $$\lim_{n=\infty} \frac{a_{n+1}}{a_n} < 1,$$

is convergent.

By condition (16) suppose that h satisfies the inequality $\lim_{n=\infty} \frac{a_{n+1}}{a_n} < h < 1$. Hence a number k exists such that for $n \geqslant k$ we have $\frac{a_{n+1}}{a_n} < h$, i.e. $a_{n+1} < a_n \cdot h$.

Thus, the series $a_k + a_{k+1} + a_{k+2} + ...$ has terms less than the terms of the geometric series $a_k + a_k \cdot h + a_k \cdot h^2 + ...$, respectively. But the last series is convergent because $0 < h < 1$. Thus, Theorem 2 implies the convergence of the series $\sum_{n=k}^{\infty} a_n$ and hence of the series $\sum_{n=1}^{\infty} a_n$, too.

4. CAUCHY CRITERION. *A series $a_1 + a_2 + ...$ with positive terms satisfying the condition*

(17) $$\lim_{n=\infty} \sqrt[n]{a_n} < 1,$$

is convergent.

Similarly as before, there exist an $h < 1$ and a k such that for $n \geqslant k$, we have $\sqrt[n]{a_n} < h$, i.e. $a_n < h^n$. Comparing the series $a_k + a_{k+1} + ...$ with the geometric series $h^k + h^{k+1} + ...$, we conclude the convergence of the first series from the convergence of the second one.

5. DIVERGENCE CRITERIA. *If the series $a_1 + a_2 + ...$ with positive terms satisfies one of the two inequalities*

(18) $$\lim_{n=\infty} \frac{a_{n+1}}{a_n} > 1 \quad \text{or} \quad \lim_{n=\infty} \sqrt[n]{a_n} > 1,$$

then the series is divergent.

---

[1] Jean le Rond d'Alembert (1717-1783), a famous French mathematician and encyclopaedist.

If the first of the inequalities (18) holds, then for sufficiently large n we have

$$\frac{a_{n+1}}{a_n} > 1, \quad \text{i.e.} \quad a_{n+1} > a_n,$$

whence the sequence $\{a_n\}$ does not converge to 0 and consequently, the series $a_1 + a_2 + \ldots$ is divergent. If the second of the inequalities (18) holds, then for sufficiently large n we have $\sqrt[n]{a_n} > 1$, i.e. $a_n > 1$ and the sequence $\{a_n\}$ is not convergent to 0, neither.

Remark. It may happen that

$$\lim_{n=\infty} \frac{a_{n+1}}{a_n} = 1 = \lim_{n=\infty} \sqrt[n]{a_n}.$$

In this case our criteria do not determine whether the given series is convergent or divergent. Then we are dealing with the so called "doubtful" case; we say also that the series does not react to the above criteria.

3.5. Applications and examples

1. *The series* $\sum_{n=1}^{\infty} \frac{c^n}{n!}$, *where* $c > 0$, *is convergent*.

Indeed,

$$\frac{a_{n+1}}{a_n} = \frac{c^{n+1}}{c^n} \cdot \frac{n!}{(n+1)!} = \frac{c}{n+1}, \quad \text{whence} \quad \lim_{n=\infty} \frac{a_{n+1}}{a_n} = 0.$$

By d'Alembert criterion we conclude that the given series is convergent.

2. *The series* $\sum_{n=1}^{\infty} \frac{n!}{n^n}$ *is convergent*.

Let $a_n = \frac{n!}{n^n}$. Hence $\frac{a_{n+1}}{a_n} = (n+1) \frac{n^n}{(n+1)^{n+1}} = \frac{1}{\left(1 + \frac{1}{n}\right)^n}$.

Thus

$$\lim_{n=\infty} \frac{a_{n+1}}{a_n} = \frac{1}{\lim_{n=\infty} \left(1 + \frac{1}{n}\right)^n} = \frac{1}{e} < 1.$$

3. $\lim_{n=\infty} \dfrac{c^n}{n!} = 0 = \lim_{n=\infty} \dfrac{n!}{n^n}$.

This merely states that the general term a_n of a convergent series converges to 0.

Defining a sequence $\{a_n\}$ divergent to ∞ *to tend more rapidly to* ∞ than a sequence $\{b_n\}$ if the condition $\lim_{n=\infty} \dfrac{b_n}{a_n} = 0$ holds, we may rewrite Formula 3 in the following way: $n!$ *tends more rapidly to* ∞ *than* c^n, *and* n^n *tends more rapidly to* ∞ *than the factorial* $n!$.

4. *The harmonic series* $1 + \dfrac{1}{2} + \dfrac{1}{3} + \ldots + \dfrac{1}{n} + \ldots$ *does not react to the d'Alembert criterion. The same holds for the series* (14).

For,
$$\lim_{n=\infty} \frac{n}{n+1} = \lim_{n=\infty} \frac{1}{\left(1 + \dfrac{1}{n}\right)} = 1.$$

As we know, the first of the above two series is divergent and the second one is convergent.

5. *The series* $\sum\limits_{n=1}^{\infty} \dfrac{\lfloor x(x-1) \cdot \ldots \cdot (x-n+1) \rfloor}{n!} c^n$ *is convergent for* $0 < c < 1$.

For, we have for non-negative and non-integer x
$$\frac{a_{n+1}}{a_n} = \frac{|x-n|}{n+1} c = \frac{\left|\dfrac{x}{n} - 1\right|}{1 + \dfrac{1}{n}} c.$$

Hence $\lim\limits_{n=\infty} \dfrac{a_{n+1}}{a_n} = c$.

The convergence of this series implies

6. $\lim\limits_{n=\infty} \dfrac{x(x-1) \cdot \ldots \cdot (x-n+1)}{n!} c^n = 0$ *for* $|c| < 1$.

7. *The series* $\sum\limits_{n=1}^{\infty} \dfrac{n^p}{c^n}$ *is convergent for* $c > 1$ (*where* p *is an integer*). *Hence* $\lim\limits_{n=\infty} \dfrac{n^p}{c^n} = 0$.

5

For, $\lim\limits_{n=\infty}\left(\dfrac{n+1}{n}\right)^p \dfrac{c^n}{c^{n+1}} = \dfrac{1}{c} < 1$.

8. The divergence of the harmonic series and the Cauchy criterion imply immediately the formula $\lim\limits_{n=\infty} \sqrt[n]{n} = 1$.

Since the sequence $\sqrt[n]{n}$ is decreasing for sufficiently large n and consists of terms > 1, so $\lim\limits_{n=\infty} \sqrt[n]{n} \geqslant 1$ exists. Supposing $\lim\limits_{n=\infty} \sqrt[n]{n} > 1$ we should get $\lim\limits_{n=\infty} \sqrt[n]{1/n} < 1$ and the Cauchy criterion would imply the convergence of the series $\sum\limits_{n=1}^{\infty} \dfrac{1}{n}$.

Remark *. The Cauchy criterion is stronger than the d'Alembert criterion. For the existence of the limit $\lim\limits_{n=\infty} \dfrac{a_{n+1}}{a_n}$ implies the existence of the limit $\lim\limits_{n=\infty} \sqrt[n]{a_n}$ (and both these limits are equal one to another, cf. § 2.11, 4). However, the last limit may exist although the first one does not exist.

3.6. Other convergence criteria

KUMMER THEOREM ([1]). *A necessary and sufficient condition for the convergence of the series $a_1 + a_2 + \ldots$ with positive terms is that a sequence of positive numbers $b_1, b_2, \ldots$ exists such that*

(19) $$\lim\limits_{n=\infty}\left(b_n \cdot \dfrac{a_n}{a_{n+1}} - b_{n+1}\right) > 0.$$

In order to prove the necessity of the condition formulated in the theorem let us assume the convergence of our series and let us write $b_n = \dfrac{s - s_n}{a_n}$, where s is the

---

([1]) Ernst Kummer, a famous German mathematician of 19th century (1810-1893), an eminent worker in theory of numbers.

sum of the series and s_n is its n-th partial sum. Then
$$b_n \cdot \frac{a_n}{a_{n+1}} - b_{n+1} = \frac{s-s_n}{a_{n+1}} - \frac{s-s_{n+1}}{a_{n+1}} = \frac{a_{n+1}}{a_{n+1}} = 1.$$
Thus, the condition (19) is satisfied.

Now, we shall prove the sufficiency of this condition. For this purpose let us assume that the inequality (19) holds. Then there exist an $h > 0$ and an index k such that the inequality
$$b_n \cdot \frac{a_n}{a_{n+1}} - b_{n+1} > h$$
holds for $n \geqslant k$.

Thus we have
$$\begin{cases} h \cdot a_{k+1} < b_k a_k - b_{k+1} a_{k+1}, \\ h \cdot a_{k+2} < b_{k+1} a_{k+1} - b_{k+2} a_{k+2}, \\ \cdots \cdots \cdots \cdots \cdots \cdots \\ h \cdot a_n < b_{n-1} a_{n-1} - b_n a_n. \end{cases}$$

Adding these inequalities we obtain
$$h(s_n - s_k) < b_k a_k - b_n a_n < b_k a_k, \quad \text{whence} \quad s_n < s_k + \frac{b_k a_k}{h}.$$

The last inequality proves that the series $a_1 + a_2 + \ldots$ is bounded. Hence this series is convergent, since it is a bounded series with positive terms.

As a corollary from Kummer theorem we obtain the following

RAABE CRITERION. *A series $a_1 + a_2 + \ldots$ with positive terms satisfying the condition*
$$(20) \qquad \lim_{n=\infty} n\left(\frac{a_n}{a_{n+1}} - 1\right) > 1$$
is convergent.

Indeed, let us write $b_n = n$. We have to prove that the condition (20) implies (19). Now,
$$\lim_{n=\infty} \left(b_n \frac{a_n}{a_{n+1}} - b_{n+1}\right) = \lim_{n=\infty} \left(n \frac{a_n}{a_{n+1}} - n - 1\right)$$
$$= \lim_{n=\infty} n\left(\frac{a_n}{a_{n+1}} - 1\right) - 1 > 0.$$

3.7. Absolutely convergent series

A series $a_1 + a_2 + ...$ is called *absolutely convergent*, if the series $|a_1| + |a_2| + ...$ is convergent. A convergent series which is not absolutely convergent is called *conditionally convergent*.

1. *If a series $a_1 + a_2 + ...$ is absolutely convergent, then it is convergent in the usual sense.*

Moreover,

(21) $$\left| \sum_{n=1}^{\infty} a_n \right| \leqslant \sum_{n=1}^{\infty} |a_n|.$$

According to the Cauchy theorem one has to estimate the expression $a_k + a_{k+1} + ... + a_n$. Now,

$$|a_k + a_{k+1} + ... + a_n| \leqslant |a_k| + |a_{k+1}| + ... + |a_n| \leqslant \sum_{i=k}^{\infty} |a_i|$$

and the last sum tends to 0 as k tends to ∞, as a rest r_{k-1} of a convergent series (cf. § 3.1). Hence for any $\varepsilon > 0$ there is a k such that $r_{k-1} < \varepsilon$, whence $|a_k + a_{k+1} + ... + a_n| < \varepsilon$ for any $n > k$.

Thus, the convergence of the series $a_1 + a_2 + ...$ is proved. At the same time, writing $s_n = a_1 + a_2 + ... + a_n$ and $t_n = |a_1| + |a_2| + ... + |a_n|$, we have $|s_n| \leqslant t_n$, and in the limit we obtain $|\lim s_n| = \lim |s_n| \leqslant \lim t_n$, i.e. the formula (21).

EXAMPLES. The geometric series $a + aq + aq^2 + ...$, where $|q| < 1$, is absolutely convergent because the series $|a| + |aq| + |aq^2| + ...$ is convergent (cf. § 3.1).

The series $\sum_{n=0}^{\infty} \frac{x^n}{n!}$ is absolutely convergent for every x (cf. § 3.5, 1; as we shall show later, its sum is equal to e^x).

The anharmonic series (7) is conditionally convergent, since the series of absolute values of its components is divergent (it is the harmonic series, cf. § 3.2 above).

Now, we shall consider the problem of the *commutativity* of infinite series. As it is well known a sum of a finite number of terms is commutative, i.e. does not depend on the arrangement of the terms. We shall show that this property remains true also for absolutely convergent series; yet it does not hold for conditionally convergent series. However, this requires that we first establish what will be understood by a rearrangement of the terms, when the number of these terms is infinite (taking into account that some terms may be repeated arbitrarily often).

By a *permutation* of the sequence of positive integers we understand a sequence of positive integers $m_1, m_2, \ldots$, such that every positive integer appears in this sequence exactly once. If $m_1, m_2, \ldots$ is a permutation of the sequence of positive integers, then we say that the series $a_{m_1} + a_{m_2} + \ldots + a_{m_n} + \ldots$ is obtained from the series $a_1 + a_2 + \ldots + a_n + \ldots$ by a rearrangement of its components.

2. *Every absolutely convergent series is commutative.*

In other words, if a series $\sum_{n=1}^{\infty} a_n$ is absolutely convergent and if $m_1, m_2, \ldots$ is a permutation of the sequence of positive integers, then

(22) $$\sum_{n=1}^{\infty} a_{m_n} = \sum_{n=1}^{\infty} a_n.$$

Let $\varepsilon > 0$. The series $|a_1| + |a_2| + \ldots$ being convergent, there exists a k such that

(23) $$\sum_{i=k+1}^{\infty} |a_i| < \varepsilon.$$

Since the sequence $\{m_n\}$ contains all positive integers, a number r exists such that the numbers $1, 2, 3, \ldots, k$ all appear among the numbers $m_1, m_2, \ldots, m_r$. But each positive integer appears in the sequence $\{m_n\}$ only once; therefore we have $m_n > k$ for each $n > r$. So by a given $n > r$, if we cancel the numbers $1, 2, \ldots, k$ in the system

$m_1, m_2, \ldots, m_r, \ldots, m_n$, then there remain in this system only numbers $> k$ (and different one from another). Thus, writing $s_n = a_1 + a_2 + \ldots + a_n$ and $t_n = a_{m_1} + a_{m_2} + \ldots + a_{m_n}$ and reducing in the difference $t_n - s_n$ the terms with equal indices, we obtain as remaining terms only terms with indices $> k$. This (and the formula for the absolute value of a sum) implies

$$|t_n - s_n| \leqslant \sum_{i=k+1}^{\infty} |a_i|, \quad \text{whence} \quad |t_n - s_n| < \varepsilon \text{ by (23)}.$$

Since the last inequality holds for each $n > r$, we obtain

$$\lim_{n=\infty} t_n = \lim_{n=\infty} s_n.$$

This is equivalent to the formula (22).

Remark. The previous theorem cannot be extended to arbitrary convergent series. For example let us rearrange the terms of the series $1 - \frac{1}{2} + \frac{1}{3} - \frac{1}{4} + \ldots$ in the following way:

$$(24) \quad 1 - \frac{1}{2} + \left(\frac{1}{3} + \frac{1}{5} + \ldots + \frac{1}{2k_1 - 1}\right) - \frac{1}{4} +$$
$$+ \left(\frac{1}{2k_1 + 1} + \frac{1}{2k_1 + 3} + \ldots + \frac{1}{2k_2 - 1}\right) - \frac{1}{6} + \ldots,$$

where the numbers $k_1, k_2, \ldots$ are chosen in such a way that

$$\frac{1}{3} + \frac{1}{5} + \ldots + \frac{1}{2k_1 - 1} > 1, \ldots,$$

$$\frac{1}{2k_n + 1} + \frac{1}{2k_n + 3} + \ldots + \frac{1}{2k_{n+1} - 1} > 1, \ldots$$

Then it is easily seen that the series (24) is divergent to ∞.

Writing $c = 1 - \frac{1}{2} + \frac{1}{3} \ldots$ (as known, $c = \log 2$) one may easily prove that

$$\tfrac{3}{2} c = c + \tfrac{1}{2} c = 1 + \tfrac{1}{3} - \tfrac{1}{2} + \tfrac{1}{5} + \tfrac{1}{7} - \tfrac{1}{4} + \tfrac{1}{9} + \tfrac{1}{11} - \tfrac{1}{6} + \ldots$$

Hence we see that by a rearrangement of terms we may obtain from the anharmonic series a divergent series or a series convergent to another limit. The following general theorem (due to Riemann [1]) holds: *given a conditionally convergent series, we may obtain a divergent series or a series convergent to a previously assigned number (finite or infinite) by a rearrangement of the terms of the given series.*

3.8. Multiplication of series

(CAUCHY) THEOREM. *Assuming the series $\sum_{n=1}^{\infty} a_n$ and $\sum_{n=1}^{\infty} b_n$ to be absolutely convergent, we have*

(25) $$\sum_{n=1}^{\infty} a_n \cdot \sum_{n=1}^{\infty} b_n = \sum_{n=1}^{\infty} c_n,$$

where

(26) $$\begin{cases} c_1 = a_1 b_1, \\ c_2 = a_1 b_2 + a_2 b_1, \\ \ldots\ldots\ldots\ldots\ldots\ldots\ldots \\ c_n = a_1 b_n + a_2 b_{n-1} + a_3 b_{n-2} + \ldots + a_n b_1, \\ \ldots\ldots\ldots\ldots\ldots\ldots\ldots \end{cases}$$

(i.e. the sum of the indices of each term in the n-th line equals $n+1$).

Let us write $s_n = a_1 + \ldots + a_n$, $t_n = b_1 + \ldots + b_n$, and $u_n = c_1 + \ldots + c_n$, i.e.

$$u_n = a_1 t_n + a_2 t_{n-1} + a_3 t_{n-2} + \ldots + a_n t_1.$$

We have to estimate the difference

(27) $s_n t_n - u_n = a_1 t_n + a_2 t_n + \ldots + a_n t_n - u_n$
$= a_2(t_n - t_{n-1}) + a_3(t_n - t_{n-2}) + \ldots + a_n(t_n - t_1).$

Since the series $\sum_{n=1}^{\infty} b_n$ and $\sum_{n=1}^{\infty} |a_n|$ are convergent and

---

[1] Bernhard Riemann (1826-1866), famous for his works in the field of geometry, the theory of analytic functions and the theory of numbers.

consequently bounded, a number M exists such that
(28) $\quad |t_j| < M \quad$ and $\quad |a_1|+|a_2|+\ldots+|a_j| < M$
for each j.

At the same time, given $\varepsilon > 0$, there exists a k such that the condition $n > m > k$ implies (cf. § 2.7, Remark (β)):
(29) $\quad |t_n - t_m| < \varepsilon \quad$ and $\quad |a_{k+1}|+|a_{k+2}|+\ldots+|a_n| < \varepsilon$.

In the following suppose that $n > 2k$. Then we have, by (27),
$$|s_n t_n - u_n| \leqslant (|a_2| \cdot |t_n - t_{n-1}| + \ldots + |a_k| \cdot |t_n - t_{n-k+1}|) + $$
$$+ (|a_{k+1}| \cdot |t_n - t_{n-k}| + \ldots + |a_n| \cdot |t_n - t_1|).$$

Applying to the first bracket formula (29) and to the second one formula (28) we obtain (taking into account that $n-k+1 > k$ and that $|t_n - t_j| \leqslant |t_n| + |t_j| < 2M$):
$$|s_n t_n - u_n| \leqslant [|a_2|+\ldots+|a_k|]\varepsilon + [|a_{k+1}|+\ldots+|a_n|] \cdot 2M$$
$$< M\varepsilon + \varepsilon \cdot 2M,$$

where the second inequality is a consequence of the second parts of the formulae (28) and (29).

So we have proved that the inequality $|s_n t_n - u_n| < 3M\varepsilon$ holds for each $n > 2k$. This means that $\lim(s_n t_n - u_n) = 0$. Since the sequences $\{s_n\}$ and $\{t_n\}$ are convergent, we have $\lim s_n t_n = \lim s_n \cdot \lim t_n$, whence we obtain $\lim s_n \lim t_n = \lim s_n t_n = \lim u_n$, i.e. the formula (25).

EXAMPLE. We shall prove that
$$\sum_{n=0}^{\infty} \frac{x^n}{n!} \cdot \sum_{n=0}^{\infty} \frac{y^n}{n!} = \sum_{n=0}^{\infty} \frac{(x+y)^n}{n!}.$$

Since both series are absolutely convergent, we may apply the formula (25). We obtain
$$\sum_{n=0}^{\infty} \frac{x^n}{n!} \cdot \sum_{n=0}^{\infty} \frac{y^n}{n!} =$$

$$= \sum_{n=0}^{\infty} \left(1 \cdot \frac{y^n}{n!} + \frac{x}{1!} \cdot \frac{y^{n-1}}{(n-1)!} + \frac{x^2}{2!} \cdot \frac{y^{n-2}}{(n-2)!} + \ldots + \frac{x^n}{n!} \cdot 1\right)$$

$$= \sum_{n=0}^{\infty} \frac{1}{n!} \left(y^n + \frac{n}{1!} xy^{n-1} + \frac{n(n-1)}{2!} x^2 y^{n-2} + \ldots + x^n\right)$$

$$= \sum_{n=0}^{\infty} \frac{(y+x)^n}{n!}$$

by the Newton binomial formula.

Let us add that the above-proved formula follows also from the equality $e^{x+y} = e^x \cdot e^y$ (cf. § 3.7).

Remarks. The theorem is also true under a weaker assumption; namely, both series have to be convergent and one of them (but not necessarily both) has to be absolutely convergent. Indeed, only the absolute convergence of the series $\sum_{n=1}^{\infty} a_n$ was employed in our proof.

Yet if both series are conditionally convergent, then the series $\sum_{n=1}^{\infty} c_n$ may be divergent. This may be shown on the following example:

$$a_n = \frac{(-1)^n}{\sqrt{n}} = b_n;$$

the series $\sum_{n=1}^{\infty} a_n$ and $\sum_{n=1}^{\infty} b_n$ are convergent, but it is easily seen that the series $\sum_{n=1}^{\infty} c_n$ is divergent.

Let us add (without proof) that if both series are conditionally convergent, then their product may be obtained by summation of the series $\sum_{n=1}^{\infty} c_n$ by the method of the "first means"; more exactly, the following theorem (due to Cesàro) holds: *if the series $a_1 + a_2 + \ldots$ and $b_1 + b_2 + \ldots$ are convergent, then*

$$\lim_{n=\infty} \frac{s_1 + s_2 + \ldots + s_n}{n} = \sum_{n=1}^{\infty} a_n \cdot \sum_{n=1}^{\infty} b_n,$$

where $s_n = c_1 + c_2 + \ldots + c_n$.

3.9*. Infinite products

Just as we consider an infinite series to be a generalization of the idea of addition to an infinite number of terms, we may consider an infinite product as another infinite operation. Here we shall restrict ourselves to give the definition and several more important properties of infinite products.

Let an infinite sequence of real numbers different from 0 be given: $a_1, a_2, \ldots, a_n, \ldots$ Let us form the sequence of products:

(30) $\quad p_1 = a_1, \; p_2 = a_1 \cdot a_2, \; \ldots, \; p_n = a_1 \cdot a_2 \cdot \ldots \cdot a_n, \; \ldots$

If the sequence (30) is convergent, then we denote its limit by the symbol

$$\prod_{n=1}^{\infty} a_n = \lim_{n=\infty} p_n$$

and we call it an *infinite product*. If this limit is finite and different from 0, then we say that the considered infinite product is *convergent*.

We call the numbers $a_1, a_2, \ldots$, the *factors* of the infinite product $a_1 \cdot a_2 \cdot \ldots$ The terms of the sequence (30) are called the *partial products*.

EXAMPLES. (α) The infinite product

$$(1+\tfrac{1}{1}) \cdot (1-\tfrac{1}{2}) \cdot (1+\tfrac{1}{3}) \cdot (1-\tfrac{1}{4}) \cdot \ldots$$

is convergent to 1. Indeed,

$$p_{2n} = \frac{2}{1} \cdot \frac{1}{2} \cdot \frac{4}{3} \cdot \frac{3}{4} \cdot \ldots \cdot \frac{2n}{2n-1} \cdot \frac{2n-1}{2n} = 1,$$

$$p_{2n+1} = p_n \cdot \frac{2n+2}{2n+1} = \frac{2n+2}{2n+1}.$$

Thus $\lim_{n=\infty} p_n = 1$.

(β) $\prod_{n=1}^{\infty} \left(1 + \frac{1}{n}\right) = \infty$, i.e. the considered infinite product is divergent to ∞, for

$$p_n = \frac{2}{1} \cdot \frac{3}{2} \cdot \ldots \cdot \frac{n+1}{n} = n+1, \quad \text{whence} \quad \lim_{n=\infty} p_n = \infty.$$

(γ) $\prod_{n=2}^{\infty} \left(1 - \frac{1}{n}\right) = 0$, i.e. the infinite product is divergent to 0, for $p_n = \frac{1}{n}$.

(δ) The infinite product $(-1) \cdot (-1) \cdot \ldots$ is divergent. Moreover, it is divergent neither to 0 nor to ∞. In this case the sequence of partial products oscillates, taking the values -1 and $+1$, alternately.

1. CAUCHY CONDITION. *A necessary and sufficient condition for the convergence of an infinite product $a_1 \cdot a_2 \cdot \ldots$ (where $a_n \neq 0$) is that to every $\varepsilon > 0$ a number k exists such that the inequality*

(31) $\qquad |a_k \cdot a_{k+1} \cdot \ldots \cdot a_n - 1| < \varepsilon$

holds for each $n > k$.

To prove the necessity of the Cauchy condition let us assume that the considered infinite product is convergent, i.e. that

(32) $\qquad \lim_{m=\infty} p_m = p \neq 0$.

Let $c = \frac{1}{2}|p|$. Then a number j exists such that $|p_m| > c$ for $m > j$.

Let a number $\varepsilon > 0$ be given. By (32), there exists a k such that $|p_n - p_{k-1}| < c\varepsilon$ for $n > k$. One may also assume that $k-1 > j$, and so $|p_{k-1}| > c$. Hence we obtain

$$\left|\frac{p_n}{p_{k-1}} - 1\right| < \frac{c\varepsilon}{|p_{k-1}|} < \varepsilon,$$

i.e. the inequality (31).

In this way we have proved that the Cauchy condition is a necessary condition for the convergence of the infinite product under consideration. Now, we shall prove the sufficiency of this condition.

Let us assume this condition to be satisfied. First of all we shall prove the sequence $\{p_n\}$ to be bounded. Substituting $\varepsilon = 1$ in the Cauchy condition we conclude that a number r exists such that for $n > r$ we have

$$|a_{r+1} \cdot a_{r+2} \cdot \ldots \cdot a_n - 1| < 1, \quad \text{i.e.} \quad \left|\frac{p_n}{p_r} - 1\right| < 1;$$

thus $|p_n - p_r| < |p_r|$, whence $|p_n| - |p_r| \leqslant |p_n - p_r| < |p_r|$ and finally $|p_n| < 2|p_r|$. Hence we conclude that the sequence (30) is bounded.

Now, let M be a number satisfying the condition $M > |p_n|$ for each positive integer n. According to (31), to any given number $\varepsilon > 0$ there exists a k such that

(33) $$\left|\frac{p_n}{p_{k-1}} - 1\right| < \varepsilon$$

for $n > k$. Hence $|p_n - p_{k-1}| < \varepsilon |p_{k-1}| < \varepsilon M$. Thus, the sequence $\{p_n\}$ is convergent. It remains to prove that its limit is $\neq 0$.

Now, supposing $\lim\limits_{n=\infty} p_n = 0$, the formula (33) (with a constant k) should give the inequality $1 \leqslant \varepsilon$ which contradicts the assumption that ε is an arbitrary positive number.

2. *If the infinite product $a_1 \cdot a_2 \cdot \ldots$ is convergent, then*

$$\lim_{n=\infty} a_n = 1.$$

Namely, let us assume that $\lim\limits_{n=\infty} p_n = p \neq 0$. Then we have $\lim\limits_{n=\infty} p_{n-1} = p$ and dividing the first equality by the second one we obtain

$$\lim_{n=\infty} a_n = \lim_{n=\infty} \frac{p_n}{p_{n-1}} = \frac{\lim\limits_{n=\infty} p_n}{\lim\limits_{n=\infty} p_{n-1}} = 1.$$

Let us note that the equation $\lim_{n=\infty} a_n = 1$ being (as we have proved) a necessary condition of the convergence of the product $\prod_{n=1}^{\infty} a_n$, is not a sufficient condition. The examples (β) and (γ) show this.

Taking into account Theorem 2 it is suitable to write the factors of an infinite product in the form $a_n = 1 + b_n$. This makes possible a simple formulation of the dependence between the convergence of the product $\prod_{n=1}^{\infty} (1 + b_n)$ and the convergence of the series $\sum_{n=1}^{\infty} b_n$. In fact, the following theorem holds:

3. *Assuming $b_n > 0$ for each n, the product $\prod_{n=1}^{\infty} (1 + b_n)$ is convergent if and only if the series $\sum_{n=1}^{\infty} b_n$ is convergent. The same concerns the product $\prod_{n=1}^{\infty} (1 - b_n)$ (assuming $b_n < 1$).*

A simple proof of this theorem may be obtained by application of the properties of the exponential function and especially of the inequality $e^x \geq 1 + x$ which we shall prove in § 8 (formula (16)). This inequality gives immediately

$$p_n \leq e^{s_n}, \quad \text{where} \quad p_n = (1 + b_1) \cdot \ldots \cdot (1 + b_n),$$

$$s_n = b_1 + \ldots + b_n.$$

Since on the other hand we have, as easily seen, $1 + s_n \leq p_n$, we conclude that the series $\{p_n\}$ and $\{s_n\}$ are either both bounded or both unbounded. Now, they are both increasing sequences; hence they are either both convergent or both divergent. Finally, the sequence $\{p_n\}$ cannot be convergent to 0, because $p_n > 1$.

Thus, the first part of our theorem is proved.

Before we proceed to the proof of the second part, let us note that if the product $\prod_{n=1}^{\infty} (1+|u_n|)$ is convergent, then the product $\prod_{n=1}^{\infty} (1+u_n)$ is convergent, too.

For let an $\varepsilon > 0$ be given and let k be chosen (according to Theorem 1) in such way that
$$(1+|u_k|) \cdot (1+|u_{k+1}|) \cdot \ldots \cdot (1+|u_n|) - 1 < \varepsilon \qquad \text{for} \qquad n > k.$$
Since, as is easily seen,
$$|(1+u_k) \cdot \ldots \cdot (1+u_n) - 1| \leqslant (1+|u_k|) \cdot \ldots \cdot (1+|u_n|) - 1,$$
so Theorem 1 implies the convergence of the product $\prod_{n=1}^{\infty} (1+u_n)$.

Now, let us assume the series $\sum_{n=1}^{\infty} b_n$ to be convergent. By the first part of our theorem the product $\prod_{n=1}^{\infty} (1+b_n)$ is also convergent and thus, by the just proved remarks, the product $\prod_{n=1}^{\infty} (1-b_n)$ is convergent, too.

There remains to prove that the convergence of the product $\prod_{n=1}^{\infty} (1-b_n)$ implies the convergence of the series $\sum_{n=1}^{\infty} b_n$. Let $q_n = (1-b_1) \cdot \ldots \cdot (1-b_n)$ and let $\lim_{n=\infty} q_n = q > 0$. Since $q \leqslant q_n \leqslant e^{-s_n}$, i.e. $e^{s_n} \leqslant \dfrac{1}{q}$, the sequence $\{s_n\}$ is bounded and so it is convergent.

An infinite product $\prod_{n=1}^{\infty} (1+u_n)$ is called *absolutely convergent*, when the product $\prod_{n=1}^{\infty} (1+|u_n|)$ is convergent. As we have just proved the absolute convergence implies convergence in the usual sense. One may also prove that—just as in the case of the series—*the value of an absolutely convergent product does not depend on the arrangement of the factors*.

Exercises on § 3

1. Prove the associativity law for convergent infinite series. Give an example of a divergent series $a_1 + a_2 + a_3 + \ldots$ such that the series $(a_1 + a_2) + (a_3 + a_4) + \ldots$ is convergent.

2. Prove that the series $\sum_{n=1}^{\infty} \frac{2^n n!}{n^n}$ is convergent and the series $\sum_{n=1}^{\infty} \frac{3^n n!}{n^n}$ is divergent.

3. Give an example of an alternating divergent series with the general term tending to 0.

4. Give an example of a convergent series $a_1 + a_2 + \ldots$ with positive term for which $\lim_{n=\infty} \frac{a_{n+1}}{a_n}$ does not exist.

5. Prove that if a series $a_1 + a_2 + \ldots$ with positive decreasing terms is convergent, then $\lim_{n=\infty} na_n = 0$.

6. Prove that the series $\sum_{n=1}^{\infty} \frac{1}{n^s}$ is convergent for $s > 1$ and divergent for $s \leqslant 1$.

Hint: apply the Raabe criterion.

7. Prove that if a series $a_1 + a_2 + \ldots$ with positive terms is convergent, then the following two series are also convergent

$$\sum_{n=1}^{\infty} \frac{\sqrt{a_n}}{n} \quad \text{and} \quad \sum_{n=1}^{\infty} \sqrt[k]{a_{n+1} \cdot a_{n+2} \cdot \ldots \cdot a_{n+k}}.$$

8. Prove that the convergence of the series $a_1 + a_2 + \ldots$ with positive decreasing terms implies the convergence of the series $\sum_{n=1}^{\infty} 2^n a_{2^n}$ (Cauchy condensation theorem).

9. Assuming the formulae

$$\sin x = \frac{x}{1!} - \frac{x^3}{3!} + \frac{x^5}{5!} - \ldots \quad \text{and} \quad \cos x = 1 - \frac{x^2}{2!} + \frac{x^4}{4!} - \ldots,$$

verify the formula $\sin(x+y) = \sin x \cos y + \sin y \cos x$ applying the theorem on multiplication of series.

10. Calculate the infinite products
$$\prod_{n=1}^{\infty} \left(1 - \frac{1}{(n+1)^2}\right), \quad \prod_{n=1}^{\infty} \left(1 + \frac{1}{n(n+2)}\right).$$

11. Examine the convergence of the product $\prod_{n=0}^{\infty} (1 + x^{2^n})$.

CHAPTER II

FUNCTIONS

§ 4. FUNCTIONS AND THEIR LIMITS

4.1. Definitions

If to any real number x there corresponds a certain number $y = f(x)$, then we say that a function f is defined over the set of real numbers.

For example by associating with any number x its square $y = x^2$ we have defined the function $f(x) = x^2$. Similarly, the equation $y = \sin x$ defines a function over the set of real numbers.

Not every function is defined over the whole set of real numbers; e.g. the equation $y = 1/x$ defines a function over the whole set of real numbers except the number 0; because this equation does not associate any number with the number 0. Similarly, the function $\sqrt{x}$ is defined only for $x \geqslant 0$; the function $\tan x$ is defined for all real numbers different from odd multiples of the number $\frac{1}{2}\pi$.

Generally, if to any x belonging to a certain set there corresponds a number $y = f(x)$, then a function is defined over this set. This set is called the set of *arguments* of the function. The numbers y of the form $y = f(x)$ are called *values* of the function f; e.g. for the function $\sqrt{x}$, the set of arguments equals the set of numbers $\geqslant 0$ and so does the set of values of this function (by the way, let us add that the symbol $\pm\sqrt{x}$ does not define a function, since it is not unique; by our definition, a function is always unique).

We shall often speak about functions defined over an interval with ends a and b. Here we shall distinguish between the *closed* interval, i.e. interval together with ends that is the set of x satisfying the condition $a \leqslant x \leqslant b$ and the *open* interval, i.e. interval without ends, that is the set of x such that $a < x < b$.

Geometrically, we interpret a function f as the set of points on the plane of the form (x, y), where $y = f(x)$. This set is called the *graph* of the function f (or the *geometric interpretation* of the function f, or the curve $y = f(x)$); e.g. the graph of the function x^2 is, as well known, the parabola and the graph of the function $1/x$, the hyperbola (cf. Fig. 2).

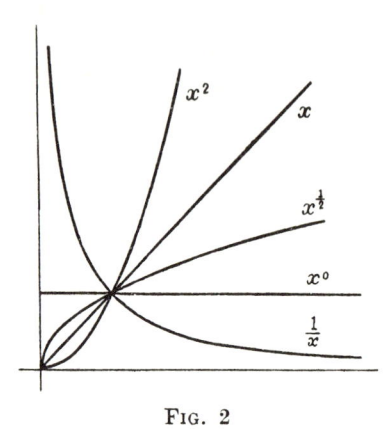

Fig. 2

Projecting the graph of a function on the x-axis we obtain the set of arguments of the function. Projecting it on the y-axis we obtain the set of values of the function. Among the sets lying on the plane, the graphs of functions are characterized by the property that no straight line parallel to the y-axis contains two points of the set.

Let us add that by our definition a function having the set of all positive integers as the set of arguments is an infinite sequence.

In all applications of mathematics, investigating relations between certain quantities, we are dealing with functions. E. g. the distance s covered by a body moving with a constant velocity v in a time t is given by the formula $s = vt$; similarly, the distance covered by a body falling under the gravitation field of the earth is a function of time, namely $s = \frac{1}{2}gt^2$.

The variable x in the expression $y = f(x)$ will be called also the *independent* variable, and y will be called the *dependent* variable. However, it has to be remembered that the constant function $y = c$ (we write also $y = $ constant) will be also reckoned among functions; its graph is a straight line parallel to the x-axis (or identical to the x-axis).

The functions about which we have spoken above are functions of one variable in distinction to functions of two and more variables. By a function of one variable, a number corresponds to any real number (belonging to the set of arguments), by a function of two variables, the value $f(x, y)$ corresponds to the values of an ordered pair x, y. Similarly, by a function of three variables, a certain number $f(x, y, z)$ corresponds to any triple of real numbers.

4.2. Monotone functions

If the condition $x < x'$ implies $f(x) < f(x')$, then the function f is said to be *strictly increasing*. If in the last inequality the sign $<$ will be replaced by $\leqslant$, then we are dealing with a function *increasing in the wider sense* (i. e. with a non-decreasing function). Similarly, if the condition $x < x'$ implies $f(x) > f(x')$ or $f(x) \geqslant f(x')$, then the function is said to be *strictly decreasing* or *decreasing in the wider sense* (non-increasing), respectively.

Increasing and decreasing functions (in narrower or wider sense) are embraced by the general name *monotone functions*.

EXAMPLES. The function x^3 is strictly increasing. The function x^2 is decreasing on the half-straight line $x \leqslant 0$ and increasing on the half-straight line $x \geqslant 0$, on the whole X-axis it is neither increasing nor decreasing.

An example of a function increasing in the wider sense (but not strictly increasing) is the function $[x]$. By this symbol we indicate the integer n satisfying the inequality $n \leqslant x$ (see Fig. 3).

The function $\sin x$ is "*piecewise monotone*": it is increasing in the interval $-\dfrac{\pi}{2} \leqslant x \leqslant \dfrac{\pi}{2}$, decreasing in the interval $\frac{\pi}{2} \leqslant x \leqslant \frac{3\pi}{2}$ etc. The functions $\cos x$, $\tan x$ possess a similar property (as do most functions which we meet in practice).

Let us note here that the above mentioned trigonometric functions have an important property of periodi-

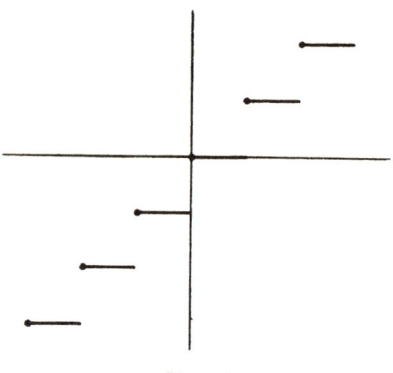

Fig. 3

city. A function $f(x)$ is called *periodic* if a number c exists (called the *period* of the function) such that the equation $f(x) = f(x+c)$ holds for every value x. The function sin has a period 2π, the function tan has a period π.

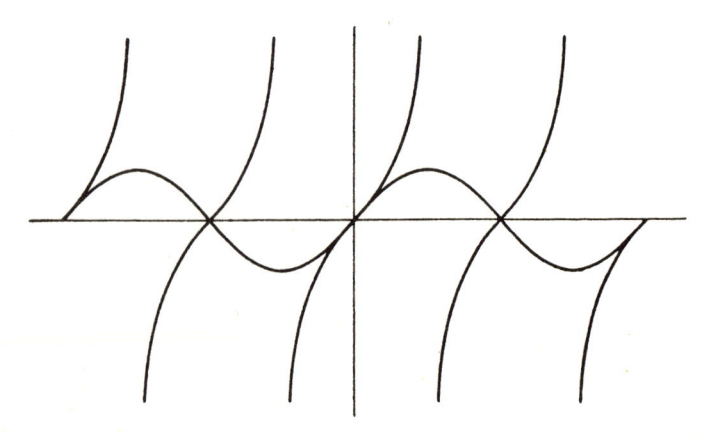

Fig. 4

Remark. In general, a function $f(x)$ will be called *piecewise monotone* in the interval $a \leqslant x \leqslant b$, if the interval $a \leqslant x \leqslant b$ may be divided by means of a finite system of points

$$a_0 < a_1 < a_2 < \ldots < a_n, \quad \text{where} \quad a_0 = a \text{ and } a_n = b$$

in intervals $a_0 a_1, a_1 a_2, \ldots, a_{n-1} a_n$ in such a way that the function $f(x)$ is defined and monotone inside of each of these intervals.

For example the function defined by the conditions: $f(x) = x$ for $0 < x < 1$, $f(1) = \frac{1}{2}$, $f(x) = x-1$ for $1 < x < 2$, is piecewise monotone in the interval $0 \leqslant x \leqslant 2$.

4.3. One-to-one functions. Inverse functions

If the function f is strictly increasing, then the equation $y = f(x)$ determines x as a function of y, i. e. this equation may be solved with respect to x (treated as an unknown). Indeed, since to different x there correspond different y, so to any y belonging to the set of values of the function f there corresponds one and only one x such that $f(x) = y$; hence x is a function of y. Thus, denoting this function by g we see that the equations

(1) $\qquad y = f(x) \quad \text{and} \quad x = g(y)$

are equivalent one to another. The function g is called *inverse to* f (sometimes we indicate it by the symbol f^{-1}).

If we should assume the function f to be strictly decreasing instead of assuming f to be strictly increasing, then we should find that f has an inverse function by the same arguments as above. For, the existence of the inverse function takes place always when the given function is one-to-one, i.e. when the condition $x \neq x'$ implies $f(x) \neq f(x')$.

EXAMPLES. The function $y = 2x+3$ has an inverse function, since solving this equation with respect to x we obtain $x = \dfrac{y-3}{2}$.

The function x^2 considered on the whole X-axis is not one-to-one because $x^2 = (-x)^2$. However, if we restrict the set of arguments to the half-straight line $x \geqslant 0$, then the function x^2 is one-to-one in this set and its inverse function is $\sqrt{x}$. Similarly, restricting the arguments to the negative half-axis we should get $-\sqrt{x}$ as the inverse function. Thus, although the function x^2 considered on the whole X-axis does not possess an inversion, it may be spoken about various "branches" of the inversion of this function (after suitable restricting of the set of its arguments).

Geometrically, a graph of a one-to-one function is characterized by the fact that on no straight line parallel to the X-axis does it contain two different points. According to the characterization of graphs of functions given in § 4.1 it is the graph of the function $x = g(y)$. To get from this function the function determined by the equation $y = g(x)$, a rotation of the graph around the straight line $y = x$ as the axis has to be performed.

Let us note the easily proved formulae

(2) $\qquad g[f(x)] = x, \quad f[g(y)] = y$.

4.4. Elementary functions

A function of the form $ax+b$, where a and b are constants, is called a *linear function*. The geometric graph of a linear function is a straight line. Conversely, any straight line, except straight lines parallel to the Y-axis, is a graph of a certain linear function. Depending on the sign of the coefficient a the function $ax+b$ is increasing (for $a > 0$) or decreasing (for $a < 0$); if $a = 0$, then the function has a constant value b. Hence our function possesses an inverse $x = \dfrac{y-b}{a}$, when $a \neq 0$.

The next (in consideration of the simplicity) class of functions is the class of *functions of the second degree*: $ax^2 + bx + c$. Geometrically they represent parabolae (when $a \neq 0$). A more general class of functions represent the

4. FUNCTIONS AND THEIR LIMITS

polynomials, i.e. functions of the form $a_0 x^n + a_1 x^{n-1} + \ldots + a_{n-1} x + a_n$. A further generalization consists of the *rational functions*, i.e. functions of the form $\dfrac{P(x)}{Q(x)}$, where $P(x)$ and $Q(x)$ are polynomials.

Polynomials are defined for all values of x. Rational functions are defined for all x except values for which the denominator vanishes, i. e. except the roots of the equation $Q(x) = 0$. In the domain of rational functions the four arithmetical operations: addition, subtraction, multiplication and division are performable; in the domain of polynomials, the first three operations only.

The function inverse to the power x^n is the *root* $\sqrt[n]{x}$. If n is odd, then this function is defined for all x and if n is even, then it is defined only for $x \geqslant 0$.

The roots are powers of the form $x^{1/n}$. More generally, we consider powers x^a with an arbitrary real exponent for $x > 0$. This function is increasing if $a > 0$ and decreasing if $a < 0$ (for $a = 0$ it is a constant) (cf. Fig. 2).

Especially, the rational function $\dfrac{1}{x}$ represents geometrically a rectangular hyperbola.

Treating the exponent in the power as a variable and the base as a constant, we get the *exponential function* a^x; we assume $a > 0$. The exponential function is increasing for $a > 1$ and decreasing for $a < 1$.

The function inverse to the exponential function is the *logarithm*: the conditions $y = \log_a x$ and $x = a^y$ are equivalent (for $0 < a \neq 1$).

In particular, the function inverse to $y = e^x$, where $e = \lim\left(1 + \dfrac{1}{n}\right)^n$ (cf. § 2.10), is the *natural logarithm* $x = \log y$ (which we write without indicating the base).

The logarithmic curves $y = \log_a x$ are obtained by rotation of the exponential curves $y = a^x$ around the straight line $y = x$.

Among the elementary functions we include also the *trigonometric functions* $\sin x$, $\cos x$, $\tan x$ and the functions inverse to the trigonometric functions (i.e. cyclometric functions): arc sin x, arc cos x, arctan x. Since the trigonometric functions are not one-to-one, in the inversion

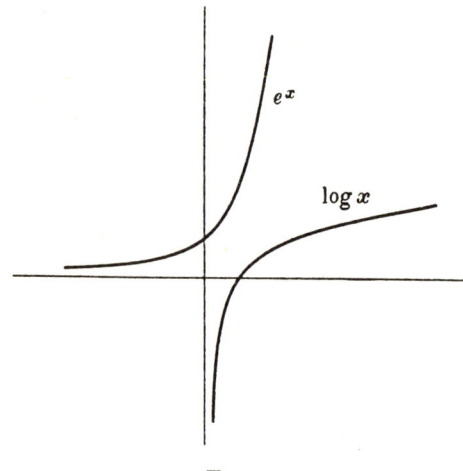

Fig. 5

of these functions we have to restrict the arguments to a suitable interval. We take $-\frac{1}{2}\pi \leqslant x \leqslant \frac{1}{2}\pi$ for $\sin x$, $0 \leqslant x \leqslant \pi$ for $\cos x$, $-\frac{1}{2}\pi < x < \frac{1}{2}\pi$ for $\tan x$. In each of

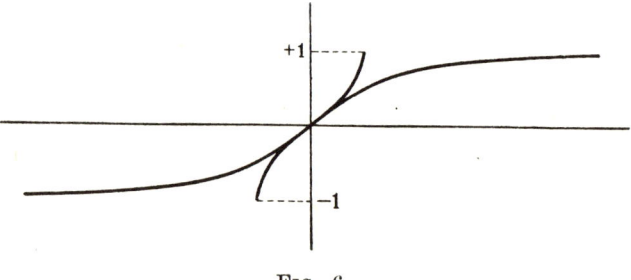

Fig. 6

these intervals the functions considered are one-to-one and so reversible, respectively. It is clear that one could take other intervals for the argument; we should get other branches of the inverse functions.

Let us note that the set of arguments of the functions arcsin$x$ and arccos$x$ is the interval $-1 \leqslant x \leqslant 1$ and the set of arguments of arctan$x$ is the whole X-axis.

Remark. In elementary courses, the method of introduction of the power a^x with a real irrational exponent is usually based implicitly on the existence of the limit $\lim\limits_{n=\infty} a^{r_n}$, where $\{r_n\}$ is a monotone sequence of rational numbers convergent to x (e.g. the sequence of decimal expansions of the number x). Now, the existence of this limit follows from the theorem on the convergence of monotone bounded sequences (§ 2.6, Theorem 3) and from the theorem on the monotony of the function a^r for rational r, known from the elementary algebra.

The equation $a^x = \lim\limits_{n=\infty} a^{r_n}$ ($a > 0$) is considered as the definition of the power for irrational exponents. Here it makes no difference which sequence of rational numbers convergent to r will be denoted by $\{r_n\}$, for $\lim\limits_{n=\infty} r'_n = \lim\limits_{n=\infty} r_n$ implies $\lim\limits_{n=\infty} a^{r'_n - r_n} = 1$ (cf. § 2.9, (ζ)), whence

$$\lim_{n=\infty} a^{r'_n} = \lim_{n=\infty} a^{r_n} \cdot \lim_{n=\infty} a^{r'_n - r_n} = \lim_{n=\infty} a^{r_n}.$$

The following formulae known for rational exponents from the elementary algebra hold also for arbitrary real exponents (when $a > 0$):

(3) $\qquad a^{x+y} = a^x \cdot a^y, \quad (ab)^x = a^x \cdot b^x, \quad (a^x)^y = a^{xy}.$

Writing $x = \lim\limits_{n=\infty} r_n$, $y = \lim\limits_{n=\infty} s_n$ and applying the above formulae to the rational exponents r_n and s_n we perform the proof by passing to the limit.

4.5. The limit of a function f at a point a

The number g is called the *limit of a function f at a point a*, if for every sequence $\{x_n\}$ convergent to a with

terms different than a the equation $\lim\limits_{n=\infty} f(x_n) = g$ holds; we denote this fact symbolically by
$$\lim\limits_{x=a} f(x) = g.$$

Thus, there exists the limit of the function f at a point a, if a common limit of all sequences $f(x_1), f(x_2), \ldots$ such that
$$\lim\limits_{n=\infty} x_n = a \quad \text{and} \quad x_n \neq a \quad \text{for each } n$$
exists.

For example $\lim\limits_{x=0} x^2 = 0$, for if $\lim\limits_{n=\infty} x_n = 0$, then $\lim\limits_{n=\infty} x_n^2 = (\lim\limits_{n=\infty} x_n)^2 = 0$.

$\lim\limits_{x=0} \cos x = 1$. Indeed, $1 - \cos x = 2\sin^2 \tfrac{1}{2}x$ and since by a known formula $|\sin t| \leqslant |t|$, we obtain $|1 - \cos x| \leqslant \tfrac{1}{2}x^2$, whence $\lim\limits_{x=0}(1-\cos x) = 0$, i.e. $\lim\limits_{x=0} \cos x = 1$.

It is also easily proved that

(4) $$\lim\limits_{x=0} a^x = 1 \quad (a > 0).$$

If a function f is defined not for all real numbers, then we assume x_n to belong to the arguments of this function; yet we do not assume this on a (assuming only that a is the limit of a certain sequence of arguments different than a); e.g. the function $\dfrac{\sin x}{x}$ is not defined for $x = 0$, however, there exists the limit of this function at the point 0, namely,

(5) $$\lim\limits_{x=0} \frac{\sin x}{x} = 1.$$

Indeed, we have

(6) $$\sin x < x < \tan x$$

for $\dfrac{\pi}{2} > x > 0$ and thus $\cos < \dfrac{\sin x}{x} < 1$ (the first of these inequalities is obtained by multiplication of the second inequality (6) by $\dfrac{\cos x}{x}$ and the second one is obtained by division of the first from the inequalities (6) by x). Since $\lim\limits_{x=0}\cos x = 1$, so we easily conclude that $\lim\limits_{n=\infty} x_n = 0$ and $x_n > 0$ imply $\lim\limits_{n=\infty} f(x_n) = 1$ (cf. §2, (17)). Finally, the assumption $x > 0$ may be replaced by $x < 0$, because $\dfrac{\sin(-x)}{-x} = \dfrac{\sin x}{x}$.

The symbol $\lim\limits_{x=a} f(x)$ will be used also to indicate the improper limit; e.g. $\lim\limits_{x=0}\dfrac{1}{x^2} = \infty$. Yet $\lim\limits_{x=0}\dfrac{1}{x}$ does not exist, since we have $\lim\limits_{n=\infty}\dfrac{1}{x_n} = \infty$ for sequences $\{x_n\}$ tending to 0 through positive values and $\lim\limits_{n=\infty}\dfrac{1}{x_n} = -\infty$ for sequences with negative terms, respectively.

However, in this case we can speak about a one-side limit at the point 0. Namely, a number g is called the *right-side* (or *left-side*) *limit* of a function f at the point a, if the conditions $\lim\limits_{n=\infty} x_n = a$ and $x_n > a$ (or $x_n < a$) imply $\lim\limits_{n=\infty} f(x_n) = g$. The symbols
$$\lim_{x=a+0} f(x) \quad \text{and} \quad \lim_{x=a-0} f(x)$$
(denoted also by $f(a+0)$ and $f(a-0)$) indicate the right- and the left-side limit of the function f at the point a, respectively.

Then we have $\lim\limits_{x=+0}\dfrac{1}{x} = \infty$, $\lim\limits_{x=-0}\dfrac{1}{x} = -\infty$. Similarly,

(7) $\lim\limits_{x=+0}\log x = -\infty, \quad \lim\limits_{x=\frac{\pi}{2}-0}\tan x = \infty, \quad \lim\limits_{x=\frac{\pi}{2}+0}\tan x = -\infty,$

$$\lim_{x=n-0}[x] = n-1.$$

However, there exist functions which do not even possess one-side limits (neither proper, nor improper).

Such is the function

(8) $$f(x) = \sin\frac{1}{x} \quad (x \neq 0).$$

It has no one-side limit at the point 0. Indeed, sin has the value 1 at the points $\frac{1}{2}\pi, \frac{1}{2}\pi+2\pi, \frac{1}{2}\pi+4\pi, \ldots, \frac{1}{2}\pi+2n\pi, \ldots$ and so writing $x_n = \dfrac{2}{(4n+1)\pi}$, we have $\lim\limits_{n=\infty} f(x_n) = 1$ and $\lim\limits_{n=\infty} x_n = 0$. At the same time, writing $x'_n = \dfrac{2}{(4n+3)\pi}$ we have $\lim\limits_{n=\infty} f(x'_n) = -1$ and $\lim\limits_{n=\infty} x'_n = 0$. Thus, no right-side limit of the function f at the point 0 exists; similarly, there exists no left-side limit at this point, too.

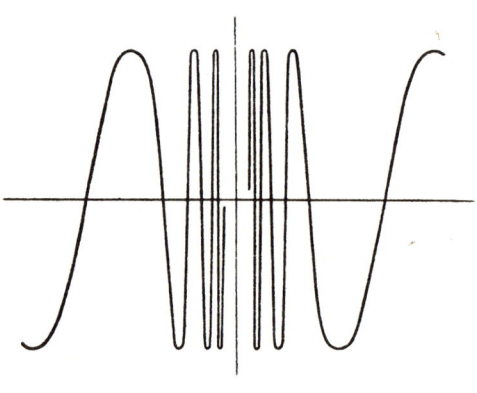

Fig. 7

Another example of a singular function constitutes the function assuming the value 1 at rational points and the value 0 at irrational points (so-called *Lejeune-Dirichlet* ([1]) *function*). It does not possess either of the one-side limits at any point.

Besides the limit of a function at a finite point a, the limit at infinity is considered, too. Namely, we denote

---

([1]) Peter Lejeune-Dirichlet (1805-1859), a famous German mathematician.

by $\lim_{x=\infty} f(x)$ the common limit of the sequences $f(x_1), f(x_2), \ldots$ such that $\lim_{n=\infty} x_n = \infty$ (if such common limit exists). For example we have

(9) $\lim_{x=\infty} \frac{1}{x} = 0, \quad \lim_{x=\infty} e^x = \infty = \lim_{x=\infty} \log x, \quad \lim_{x=-\infty} e^x = 0.$

The trigonometric functions sin, cos, tan do not possess any limit at $\pm\infty$.

4.6. Operations on the limit

Assuming the existence (and finiteness) of the limits $\lim_{x=a} f(x)$ and $\lim_{x=a} g(x)$, we obtain the formulae:

(10) $\lim_{x=a}[f(x)+g(x)] = \lim_{x=a} f(x) + \lim_{x=a} g(x),$

(11) $\lim_{x=a}[f(x)-g(x)] = \lim_{x=a} f(x) - \lim_{x=a} g(x),$

(12) $\lim_{x=a}[f(x) \cdot g(x)] = \lim_{x=a} f(x) \cdot \lim_{x=a} g(x),$

(13) $\lim_{x=a} \frac{f(x)}{g(x)} = \frac{\lim_{x=a} f(x)}{\lim_{x=a} g(x)}, \quad when \quad \lim_{x=a} g(x) \neq 0.$

The above formulae remain true if a is replaced by ∞ or $-\infty$. They remain also true for one-side limits.

These formulae are consequences of suitable formulae for infinite sequences (§ 2, (6)-(9)). For example we shall prove the formula (13).

Let $\lim_{x=a} f(x) = A$, $\lim_{x=a} g(x) = B \neq 0$ and let $\lim_{x=a} x_n = a$, where $x_n \neq a$. Then we have $\lim_{n=\infty} f(x_n) = A$, $\lim_{n=\infty} g(x_n) = B$. Hence we conclude from the formula for the quotient of limits (§ 2, (9)) that

$$\lim_{n=\infty} \frac{f(x_n)}{g(x_n)} = \frac{\lim_{n=\infty} f(x_n)}{\lim_{n=\infty} g(x_n)},$$

which implies formula (13).

Similarly we conclude from formulae (15) and (17) § 2 that, if the limits $\lim\limits_{x=a} f(x)$ and $\lim\limits_{x=a} g(x)$ exist, then

(14) *the inequality* $f(x) \leqslant g(x)$ *implies* $\lim\limits_{x=a} f(x) \leqslant \lim\limits_{x=a} g(x)$,

(15) *the inequality* $f(x) \leqslant h(x) \leqslant g(x)$ *together with* $\lim\limits_{x=a} f(x) = \lim\limits_{x=a} g(x)$ *imply* $\lim\limits_{x=a} h(x) = \lim\limits_{x=a} f(x) = \lim\limits_{x=a} g(x)$.

Here the formulae are valid for $a = \pm\infty$ and for one-side limits.

The following formula on the composition of functions holds:

(16) *If* $\lim\limits_{x=a} f(x) = A$ *and* $\lim\limits_{y=A} g(y) = B$, *then* $\lim\limits_{x=a} g[f(x)] = B$,

if the function f does not assume the value A in a neighbourhood of the point a (i.e. in a certain open interval containing a).

This formula remains true also for infinite values A, B and for one-side limits, after accepting to write $\lim\limits_{x=a} f(x) = A+0$, if $\lim\limits_{x=a} f(x) = A$ and if $f(x) > A$ in a neighbourhood of a (an analogous convention regards the symbol $A-0$ and the one-side limit).

Namely, let $\lim\limits_{n=\infty} x_n = a$ and $x_n \neq a$. Then we have $\lim\limits_{n=\infty} f(x_n) = A$. Let us write $y_n = f(x_n)$. Since the function f does not assume the value A in a certain interval containing the point a inside, so we have, for sufficiently large n, $f(x_n) \neq A$, i.e. $y_n \neq A$. Thus $\lim\limits_{n=\infty} g(y_n) = B$, i.e. $\lim\limits_{n=\infty} g[f(x_n)] = B$. Hence we conclude that $\lim\limits_{x=a} g[f(x)] = B$.

In particular we have the following formula which enables us to reduce the calculation of limits at ∞ to calculation of limits at finite points:

(17) *If* $\lim\limits_{x=+0} f\left(\dfrac{1}{x}\right)$ *exists, then* $\lim\limits_{t=\infty} f(t) = \lim\limits_{x=+0} f\left(\dfrac{1}{x}\right)$.

Indeed, substituting $x = \frac{1}{t}$ we have $\lim\limits_{t=\infty} \frac{1}{t} = +0$, whence we get the formula (17) according to the previous one (writing $a = \infty$ and $A = +0$).

As is easily seen, the reverse relation also holds:

(18) *If* $\lim\limits_{t=\infty} f(t)$ *exists, then* $\lim\limits_{x=+0} f\left(\frac{1}{x}\right) = \lim\limits_{t=\infty} f(t)$.

EXAMPLES. (α) The following formula has an important application in the differential calculus:

$$\lim_{x=0} (1+x)^{\frac{1}{x}} = e. \tag{19}$$

To deduce this formula we shall consider two following cases:

1) $x > 0$. Let us write $y = \frac{1}{x}$. One has to prove that

$$\lim_{y=\infty} \left(1 + \frac{1}{y}\right)^y = e \tag{20}$$

according to the formula (18). Now, let $\lim\limits_{n=\infty} y_n = \infty$ and let us write $k_n = [y_n]$, i.e. that k_n is a positive integer satisfying the double-inequality

(21) $\qquad k_n \leqslant y_n < k_n + 1$.

Hence

$$\frac{1}{k_n+1} \leqslant \frac{1}{y_n} \leqslant \frac{1}{k_n} \quad \text{(for } y_n \geqslant 1\text{)}$$

and thus

$$\left(1+\frac{1}{k_n+1}\right)^{k_n} \leqslant \left(1+\frac{1}{y_n}\right)^{y_n} \leqslant \left(1+\frac{1}{k_n}\right)^{k_n+1},$$

i.e.

$$\left(1+\frac{1}{k_n+1}\right)^{k_n+1} \cdot \frac{1}{1+\frac{1}{k_n+1}} \leqslant \left(1+\frac{1}{y_n}\right)^{y_n} \leqslant \left(1+\frac{1}{k_n}\right)^{k_n} \cdot \left(1+\frac{1}{k_n}\right).$$

Since $\lim\limits_{n=\infty} y_n = \infty$ and $k_n > y_n - 1$, we have $\lim\limits_{n=\infty} k_n = \infty$ and since $\lim\limits_{n=\infty} \left(1 + \dfrac{1}{n}\right)^n = e$, we conclude (cf. § 2, Exercise 5) that $\lim\limits_{n=\infty} \left(1 + \dfrac{1}{k_n}\right)^{k_n} = e$. Finally, since

$$\lim_{n=\infty} \frac{1}{1 + \dfrac{1}{k_n + 1}} = 1 = \lim_{n=\infty} \left(1 + \frac{1}{k_n}\right),$$

we obtain

$$\lim_{n=\infty} \left(1 + \frac{1}{y_n}\right)^{y_n} = e$$

according to the formula of the double-inequality (cf. § 2, (17)).

So the formula (20) and thus the formula (19), too, are proved for $x > 0$.

2) Let $x < 0$. Writing $y = \dfrac{1}{x}$, we have

$$\lim_{x=-0}(1+x)^{\frac{1}{x}} = \lim_{x=+\infty}(1-x)^{-\frac{1}{x}} = \frac{1}{\lim\limits_{y=\infty}\left(1-\dfrac{1}{y}\right)^y}.$$

Hence it remains to prove that

(22) $$\lim_{y=\infty}\left(1 - \frac{1}{y}\right)^y = \frac{1}{e}.$$

Let $\lim\limits_{n=\infty} y_n = \infty$. The formula (21) gives

$$\left(1 - \frac{1}{k_n}\right)^{k_n+1} \leqslant \left(1 - \frac{1}{y_n}\right)^{k_n+1} \leqslant \left(1 - \frac{1}{y_n}\right)^{y_n}$$

$$\leqslant \left(1 - \frac{1}{k_n+1}\right)^{y_n} \leqslant \left(1 - \frac{1}{k_n+1}\right)^{k_n}.$$

However, $\lim\limits_{n=\infty}\left(1 - \dfrac{1}{n}\right)^n = \dfrac{1}{e}$ (cf. § 2.10, (26)); hence reasoning as before we obtain $\lim\limits_{n=\infty}\left(1 - \dfrac{1}{y_n}\right)^{y_n} = \dfrac{1}{e}$. Thus the formula (22) is proved.

(β) $\lim\limits_{x=\infty} \dfrac{e^x}{x} = \infty$, i.e. the function e^x increases more rapidly than x.

Namely, let $\lim\limits_{n=\infty} x_n = \infty$ and let $k_n = [x_n]$ as before. Thus,

$$\frac{e^{x_n}}{x_n} \geqslant \frac{e^{k_n}}{k_n+1} \geqslant \frac{e^{k_n}}{2k_n} \quad \text{for} \quad x_n > 1.$$

Since (§ 3.5, 7) $\lim\limits_{n=\infty} \dfrac{e^n}{n} = \infty$ and $\lim\limits_{n=\infty} k_n = \infty$, so $\lim\limits_{n=\infty} \dfrac{e^{k_n}}{k_n} = \infty$

and thus $\lim\limits_{n=\infty} \dfrac{e^{x_n}}{x_n} = \infty$. Whence the required formula follows.

(γ) $\lim\limits_{x=\infty} \dfrac{x}{\log x} = \infty$.

Since $x = e^{\log x}$ and $\lim\limits_{x=\infty} \log x = \infty$, so substituting $y = \log x$ we obtain (cf. (16)):

$$\lim\limits_{x=\infty} \frac{x}{\log x} = \lim\limits_{x=\infty} \frac{e^{\log x}}{\log x} = \lim\limits_{y=\infty} \frac{e^y}{y} = \infty.$$

(δ) $\lim\limits_{x=+0} (x \log x) = 0$.

For

$$\lim\limits_{x=+0}(x\log x) = \lim\limits_{t=\infty} \frac{\log \dfrac{1}{t}}{t} = -\lim\limits_{t=\infty}\frac{\log t}{t}= 0.$$

4.7. Conditions for the existence of the limit

The following theorem corresponds to the theorem on the convergence of monotone bounded sequences:

THEOREM 1. *If a function f is increasing (in the wider sense) and bounded above, then $\lim\limits_{x=a-0} f(x)$ exists at every point a.*

An analogous theorem holds for decreasing functions.

By a function *bounded above* (or *below*) we understand a function, the set of values of which is bounded above (below), i.e. a function f such that a number M exists for which the inequality $M > f(x)$ (or $M < f(x)$) holds for every x.

The proof of Theorem 1. The sequence $\left\{a - \dfrac{1}{n}\right\}$ being increasing, the sequence $\left\{f\left(a - \dfrac{1}{n}\right)\right\}$ is increasing, too. But since the last sequence is also bounded, hence it is convergent. Let $\lim\limits_{n=\infty} f\left(a - \dfrac{1}{n}\right) = g$. It remains for us to prove that the conditions $\lim\limits_{n=\infty} x_n = a$ and $x_n < a$ imply the equation

$$\lim_{n=\infty} f(x_n) = g.$$

Let an $\varepsilon > 0$ be given. Then there is an m such that $g - f\left(a - \dfrac{1}{m}\right) < \varepsilon$. To this m we choose a number k such that the inequality $n > k$ implies $a - \dfrac{1}{m} < x_n$. Hence $f\left(a - \dfrac{1}{m}\right) < f(x_n)$, whence

$$g - f(x_n) < g - f\left(a - \dfrac{1}{m}\right) < \varepsilon.$$

Moreover, to each n there exists an r_n such that $x_n < a - \dfrac{1}{r_n}$. Hence

$$f(x_n) < f\left(a - \dfrac{1}{r_n}\right) \leqslant g, \quad \text{i. e.} \quad g - f(x_n) > 0.$$

Hence we conclude that $\lim\limits_{n=\infty} f(x_n) = g$.

Remark. Theorem 1 may be generalized to the case of unbounded monotone functions just as to the case $a = \pm\infty$. It may be written also in the following, slightly more general, form:

If a function f is monotone (in the wider sense) and bounded, then the limits $\lim\limits_{x=a\pm 0} f(x)$ *exist at every point a.*

4. FUNCTIONS AND THEIR LIMITS

The proof is completely analogous.

THEOREM 2. *If a function f has no (finite) limit at a point a, then a sequence $\{x_n\}$ exists such that $\lim\limits_{n=\infty} x_n = a$, $x_n \neq a$ and that the sequence $\{f(x_n)\}$ is divergent.*

For, let us suppose the sequence $\{f(x_n)\}$ to be convergent for every such sequence $\{x_n\}$. Since the function f has no limit at the point a, so two sequences $\{x_n\}$ and $\{x'_n\}$ must exist satisfying our conditions and such that $\lim\limits_{n=\infty} f(x_n) \neq \lim\limits_{n=\infty} f(x'_n)$. Let us consider the sequence $x_1, x'_1, x_2, x'_2, \ldots, x_n, x'_n, \ldots$ This sequence is convergent to a and all its terms are $\neq a$, but the sequence $f(x_1), f(x'_1), f(x_2), f(x'_2), \ldots$ is divergent.

The following theorem includes the so-called *Cauchy definition* of the limit of a function at a point:

THEOREM 3. *A necessary and sufficient condition for the equation $\lim\limits_{x=a} f(x) = g$ to be valid is that to every $\varepsilon > 0$ a number $\delta > 0$ exists such that the inequality $0 < |x-a| < \delta$ implies $|f(x)-g| < \varepsilon$.*

First, let us suppose that the Cauchy condition is not satisfied, i.e. that there exists an $\varepsilon > 0$ such that to any $\delta > 0$ an x exists for which $0 < |x-a| < \delta$, but $|f(x)-g| \geqslant \varepsilon$. Especially, substituting $\delta = \dfrac{1}{n}$ we conclude that there exists a sequence $\{x_n\}$ for which

$$0 < |x_n - a| < \frac{1}{n} \tag{23}$$

and

$$|f(x_n) - g| \geqslant \varepsilon. \tag{24}$$

It follows from (23) that $\lim\limits_{n=\infty} x_n = a$ and $x_n \neq a$. Hence, supposing $\lim\limits_{x=a} f(x) = g$ we should have $\lim\limits_{n=\infty} f(x_n) = g$. But the last formula is contradictory to the inequality (24).

Thus we have proved the Cauchy condition to be a necessary condition for the validity of the equation

$\lim_{x=a} f(x) = g$. We shall now prove this condition to be a sufficient one.

Let an $\varepsilon > 0$ be given and let $\lim_{n=\infty} x_n = a$ and $x_n \neq a$. Then by the Cauchy condition, a number $\delta > 0$ exists such that the inequality $0 < |x_n - a| < \delta$ implies $|f(x_n) - g| < \varepsilon$. But in consideration of the equation $\lim_{n=\infty} x_n = a$, the inequality $|x_n - a| < \delta$ holds for all n beginning with a certain k. For the same n we have the inequality $|f(x_n) - g| < \varepsilon$. This means that $\lim_{n=\infty} f(x_n) = g$. Hence we conclude that $\lim_{x=a} f(x) = g$.

The following theorem corresponds to the Cauchy theorem in the theory of sequences (§ 2.7):

THEOREM 4. *A necessary and sufficient condition for the existence of the (finite) limit of a function f at a point a is that to any* $\varepsilon > 0$ *a number* $\delta > 0$ *exists such that the conditions*

(25) $\qquad 0 < |x - a| < \delta \quad$ and $\quad 0 < |x' - a| < \delta$

imply $|f(x) - f(x')| < \varepsilon$.

Indeed, our condition is necessary, for if $\lim_{x=a} f(x) = g$, then to any given $\varepsilon > 0$ there is a $\delta > 0$ such that the condition $0 < |x - a| < \delta$ implies $|f(x) - g| < \tfrac{1}{2}\varepsilon$. Hence, if the conditions (25) are satisfied, then the inequalities $|f(x) - g| < \tfrac{1}{2}\varepsilon$ and $|f(x') - g| < \tfrac{1}{2}\varepsilon$ hold. Adding both these inequalities we obtain $|f(x) - f(x')| < \varepsilon$.

The condition is sufficient. Indeed, let us suppose that the limit of the function f at the point a does not exist. Then there exists by Theorem 2 a sequence $\{x_n\}$ such that $\lim_{n=\infty} x_n = a$, $x_n \neq a$. Suppose the condition formulated in the theorem is satisfied. Since the equation $\lim_{n=\infty} x_n = a$ implies the existence of a k such that for $n \geqslant k$, $x = x_n$ and $x' = x_k$ may be substituted in the inequalities (25), so we have $|f(x_n) - f(x_k)| < \varepsilon$. By the

Cauchy theorem on sequences, we conclude the convergence of the sequence $\{f(x_n)\}$, in contradiction to our assumption.

Remark. For $a = \infty$ Theorems 3 and 4 may be formulated as follows:

THEOREM 3'. *A necessary and sufficient condition for the equation* $\lim_{x=\infty} f(x) = g$ *is that to any* $\varepsilon > 0$ *a number* r *exists such that the inequality* $x > r$ *implies* $|f(x) - g| < \varepsilon$.

THEOREM 4'. *A necessary and sufficient condition for the existence of the (finite) limit* $\lim_{n=\infty} f(x)$ *is that to any* $\varepsilon > 0$ *a number* r *exists such that the conditions* $x > r$ *and* $x' > r$ *imply* $|f(x) - f(x')| < \varepsilon$.

The proofs of these theorems are completely analogous to the proofs of Theorems 3 and 4.

Exercises on § 4

1. Sketch the graphs of the functions

$$\frac{x}{x+1}, \quad x-[x], \quad x\sin\frac{1}{x}, \quad x^2\sin\frac{1}{x}, \quad \frac{1}{x}\sin\frac{1}{x}$$

and calculate the limits of these functions at the point 0.

2. Find the limits of a polynomial at ∞, i. e. $\lim_{x=\pm\infty} (x^n + a_{n-1}x^{n-1} + \ldots + a_1 x + a_0)$ (consider the case of n even and the case of n odd, seperately).

3. Evaluate:

$$\lim_{x=\infty} \frac{\sin ax}{x}, \quad \lim_{x=1} \frac{x^3-1}{x^2-1}.$$

4. Let a bounded function $f(x)$ be given in the interval $a \leqslant x \leqslant b$. Let us divide this interval into a finite number of intervals by means of the points $a_0 < a_1 < \ldots < a_n$ (where $a_0 = a$, $a_n = b$) and let us consider the sum

$$s = |f(a_1) - f(a_0)| + |f(a_2) - f(a_1)| + \ldots + |f(a_n) - f(a_{n-1})|.$$

If the set of the numbers s corresponding to all possible partitions of the interval ab is bounded, then the function f is said to be of *bounded variation*.

Prove that the sum of two functions of bounded variation is a function of bounded variation; a monotone function is of bounded variation.

5. Prove that if a function f bounded in each finite interval satisfies the condition $\lim_{x=\infty} [f(x+1)-f(x)] = g$, then $\lim_{x=\infty} \frac{f(x)}{x} = g$.

6. Prove that the formula $f(x) = \lim_{n=\infty} \lim_{k=\infty} (\cos n!\, \pi x)^{2k}$ represents the Dirichlet function considered in § 4.5.

7. A function $f(x)$ is called an *even function*, if the condition $f(x) = f(-x)$ holds for every x. Analogously, if the condition $f(x) = -f(-x)$ is satisfied, then the function f is called an *odd function*.

Give examples of even functions and odd functions, among the trigonometric functions.

Characterize even and odd functions by means of geometrical conditions.

8. Prove that every function is a sum of an even function and an odd function.

(Note that the function $g(x) = f(x) + f(-x)$ is even.)

§ 5. CONTINUOUS FUNCTIONS

5.1. Definition

A function f is said to be *continuous at a point a*, if the condition

(1) $$f(a) = \lim_{x=a} f(x)$$

is satisfied.

According to the definition of the symbol $\lim_{x=a} f(x)$ this means that the equation $\lim_{n=\infty} x_n = a$ implies the equation $\lim_{n=\infty} f(x_n) = f(a)$. (Obviously, assumption $x_n \neq a$ may be omitted here.)

Thus, if a function f is continuous at a point a, then

(2) $\quad f(\lim_{n=\infty} x_n) = \lim_{n=\infty} f(x_n)$, whenever $\lim_{n=\infty} x_n = a$.

If in the equation (1) we replace the limit by a one-side limit, we obtain the definition of the *one-side continuity*. Namely, a function f is right-side continuous (or left-side continuous) at a point a, if

$$f(a) = \lim_{x=a+0} f(x) \quad (\text{or } f(a) = \lim_{x=a-0} f(x)).$$

If a function f is defined not for all x, then we restrict x obviously to the set of arguments of the function $f(x)$; e.g. if the set of arguments of a function f is the closed interval $a \leqslant x \leqslant b$, then the continuity of the function f at the point a means the same as its right-side continuity.

A function continuous at every value of the argument is called briefly a *continuous function*. In particular, a continuous function defined in the interval $a \leqslant x \leqslant b$ is a function continuous at every point inside this interval, right-side continuous at a and left-side continuous at b.

A function f will be said also to be *piecewise continuous in the interval* $a \leqslant x \leqslant b$, if this interval may be divided by means of a finite system of points $a_0 < a_1 < a_2 < ... < a_n$, where $a_0 = a$ and $a_n = b$, into intervals $a_{k-1}a_k$, $k = 1, 2, ..., n$ in such a way that the function f is continuous inside of each of these intervals and that the limits $\lim_{x=a_{k-1}+0} f(x)$ and $\lim_{x=a_k-0} f(x)$ exist; i.e. if for each k the function $f_k(x)$ defined by the conditions $f_k(x) = f(x)$ for $a_{k-1} < x < a_k$ and $f_k(a_{k-1}) = f(a_{k-1}+0)$, $f_k(a_k) = f(a_k-0)$, is continuous in the interval $a_{k-1} \leqslant x \leqslant a_k$.

Thus, a function having a finite number of points of discontinuity in which both the one-side limits exist is piecewise continuous.

EXAMPLES. As we have proved in § 4.5, $\lim_{x=0} x^2 = 0$. This means that the function x^2 is continuous at the point 0 (though it is easily seen that it is a function continuous at every point). Similarly, the equation $\lim_{x=0} \cos x = 1 = \cos 0$ implies the continuity of the function $\cos x$ at the point 0.

The function $[x]$ is discontinuous at integer points; speaking more precisely: it is right-side continuous, but not left-side continuous at these points.

The function $\sin \frac{1}{x}$ is discontinuous at the point 0 independently of the value that could be given to this function at this point; for its limit at the point 0 does not exist. The Dirichlet function (cf. § 4.5) is discontinuous at every point.

The notion of continuity may be extended also to the points at the infinity, understanding by a function continuous at ∞ a function f such that the limit $\lim_{x=\infty} f(x)$ exists.

5.2. Cauchy characterization of continuity. Geometrical interpretation

The following theorem may be considered as the definition of continuity (the "Cauchy definition" in distinction to the "Heine definition" which we have introduced in § 5.1). It follows directly from § 4.7, Theorem 3.

A necessary and sufficient condition for a function f to be continuous at a point a is that for any $\varepsilon > 0$ there exists a number $\delta > 0$ such that the inequality $|x-a| < \delta$ implies $|f(x)-f(a)| < \varepsilon$.

Speaking more figuratively, this means that to sufficiently small increments of the independent variable there correspond arbitrarily small increments of the dependent variable. This may be written as follows:

(3) *the condition* $|h| < \delta$ *implies* $|f(a+h)-f(a)| < \varepsilon$.

5. CONTINUOUS FUNCTIONS

Geometrically, the continuity of a function f at a point a means as follows: let P denote a strip parallel to the X-axis and containing the straight line $y = f(a)$ inside; then there exists a strip Q parallel to the Y-axis and containing the straight line $x = a$ inside, such that the part of the curve given by the equation $y = f(x)$ included in the strip Q, is also included in the strip P.

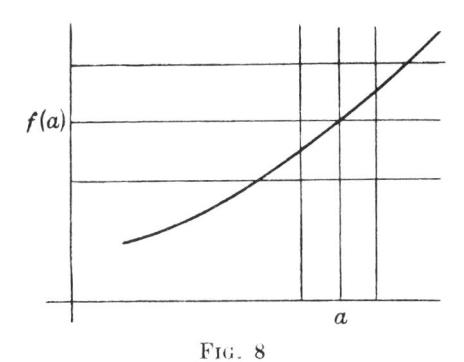

Fig. 8

We may pass from the analytic definition based on the Cauchy condition to the above geometrical definition immediately, assuming the strip P to be defined by the straight lines $y = f(a) \pm \varepsilon$ and the strip Q to be defined by the straight lines $x = a \pm \delta$.

It is easily seen from the above definitions that the continuity of a function f at a point a is a *local* property in the sense that it depends on the behaviour of the function in the neighbourhood of the point a. In order to find whether a function f is continuous at a point a it suffices to know this function in an arbitrarily small interval containing the point a inside.

5.3. Continuity of elementary functions

It follows immediately from the formulae on the operations on limits of functions (§ 4, (10)-(13)) that *the arithmetical operations on continuous functions give continuous functions,* again. Speaking more exactly, *if the* *functions f and g are continuous at a point a, then the sum, the difference, the product and the quotient (when $g(a) \neq 0$) of these functions are continuous at the point a, also.*

For example assuming $\lim_{x=a} f(x) = f(a)$ and $\lim_{x=a} g(x) = g(a)$ we have

$$\lim_{x=a}[f(x) \cdot g(x)] = \lim_{x=a} f(x) \cdot \lim_{x=a} g(x) = f(a) \cdot g(a)$$

which means that the function $f(x) \cdot g(x)$ is continuous at the point a.

Hence it follows that *a polynomial is a continuous function.*

Namely, the constant function $f(x) = $ const. and the function $f(x) = x$ (the identity) are continuous functions, as easily seen.

More generally, *any rational function, i.e. function of the form* $f(x) = \dfrac{P(x)}{Q(x)}$, *where $P(x)$ and $Q(x)$ are polynomials, is a continuous function* (obviously at such points at which it is defined, i.e. everywhere except of the roots of the denominator).

Before proceeding to the proof of the continuity of other elementary functions, let us note that according to the formula (1), the continuity of a function f at a point x means that $\lim_{h=0} f(x+h) = f(x)$, i.e. that

(4) $$\lim_{h=0}[f(x+h) - f(x)] = 0.$$

We shall prove now that *the exponential function a^x ($a > 0$) is continuous.*

Indeed, we have $a^{x+h} - a^x = a^x(a^h - 1)$ and $\lim_{h=0} a^h = 1$ (cf. § 4, (4)). Hence $\lim_{h=0}(a^{x+h} - a^x) = 0$ and this means according to the formula (4) that the function a^x is continuous.

The trigonometric functions sin *and* cos *are continuous.*

Indeed,

$$\sin(x+h) - \sin x = 2\sin\frac{h}{2} \cdot \cos\left(x + \frac{h}{2}\right)$$

and since $|\sin t| \leqslant |t|$ and $|\cos t| \leqslant 1$, we have
$|\sin(x+h) - \sin x| \leqslant |h|$, whence $\lim_{h=0} |\sin(x+h) - \sin x| = 0$,
i.e.
$$\lim_{h=0} \sin(x+h) - \sin x = 0.$$

Similarly,
$$|\cos(x+h) - \cos x| = 2\left|\sin\frac{h}{2}\right| \cdot \left|\sin\left(x+\frac{h}{2}\right)\right| \leqslant |h|,$$
whence
$$\lim_{h=0} \cos(x+h) - \cos x = 0.$$

Since $\tan x = \dfrac{\sin x}{\cos x}$, the function $\tan x$ is a continuous function as a quotient of continuous functions. The functions
$$\sec x = \frac{1}{\cos x}, \quad \cot x = \frac{\cos x}{\sin x}, \quad \cosec x = \frac{1}{\sin x}$$
are also continuous functions (at every x at which they are defined).

Among the elementary functions there remain still to consider the functions inverse to the above considered, namely, the logarithm and the cyclometric functions. However, we shall not prove the continuity of these functions now, because it follows from a general theorem on the continuity of functions inverse to continuous functions which will be proved something later.

However, we shall prove now that *the composition of continuous functions is a continuous function*. Speaking more exactly, *if two functions $y = f(x)$ and $z = g(y)$ are given and if the function f is continuous at a point a and if the function g is continuous at the point $f(a)$, then the function $g(f(x))$ is continuous at the point a.*

For, let $\lim_{n=\infty} x_n = a$. Hence $\lim_{n=\infty} f(x_n) = f(a)$. Writing $b = f(a)$ and $y_n = f(x_n)$, we have then $\lim_{n=\infty} y_n = b$, whence the continuity of the function g implies

$$\lim_{n=\infty} g(y_n) = g(b), \quad \text{i.e.} \quad \lim_{n=\infty} g(f(x_n)) = g(f(a)).$$

But this means the continuity of the composite function $g(f(x))$ at the point a.

Hence it follows that taking the elementary functions considered in § 4.4 as the starting-point and applying to these functions the arithmetical operations and the composition, we yet remain in the domain of continuous functions.

For example the function $x + \tan(x^2)$ is a continuous function. Similarly, the function $\sin \frac{1}{x}$ is a continuous function in the set of its arguments, i.e. for $x \neq 0$. But the function f defined by the conditions:

(5) $\quad f(x) = \sin \dfrac{1}{x} \quad$ for $\quad x \neq 0 \quad$ and $\quad f(0) = 0$

is discontinuous at the point 0. As we know, the limit of the function f at the point 0 does not exist at all (cf. § 4.5, (8)).

Hence it follows also that the function defined by the conditions (5) cannot be written by means of continuous functions only. Similarly, the function defined as follows:

$$f(x) = 1 \quad \text{for} \quad x \neq 0 \quad \text{and} \quad f(0) = 0$$

is discontinuous at the point 0. However, the last function may be written by means of a limit. Namely,

$$f(x) = \lim_{t=\infty} \frac{xt}{1+xt} = \lim_{n=\infty} \frac{nx}{1+nx}.$$

Remark. The following formulation of the theorem on the composition of functions—a little more general than the above—is convenient in applications: *if* $\lim_{x=a\pm 0} f(x)$ *exists and if* g *is continuous at this point, then*

(6) $\qquad\qquad \lim\limits_{x=a\pm 0} g[f(x)] = g[\lim\limits_{x=a\pm 0} f(x)].$

The proof remains almost without change.

EXAMPLE. $\lim\limits_{x=+0} x^x = 1$, for

$$\lim_{x=+0} x^x = \lim_{x=+0} e^{x\log x} = e^{\lim\limits_{x=+0} (x\log x)} = e^0 = 1.$$

Here $f(x) = x\log x$, $g(x) = e^x$. The function f is not defined at the point 0; however, it has a right-side limit ($= 0$; cf. § 4.6, (δ)) at this point.

5.4. General properties of continuous functions

1. THEOREM ON THE UNIFORM CONTINUITY. *If a function f is continuous in a closed interval $a \leqslant x \leqslant b$, then to every $\varepsilon > 0$ there exists a $\delta > 0$ such that the inequality $|x-x'| < \delta$ implies $|f(x)-f(x')| < \varepsilon$.*

Before proceeding to the proof, let us note that the number δ mentioned in the above theorem does not depend on x in distinction to the case of the definition of a function continuous at every point x (belonging to the considered set of arguments), where such dependence holds. This independence of the number δ of the variable x we emphasize saying "uniform" continuity.

The proof of the theorem will be given by *reductio ad absurdum*. Let us suppose that a number $\varepsilon > 0$ exists such that, given any $\delta > 0$, there exists a pair of arguments x and x' such that $|x-x'| < \delta$ and $|f(x)-f(x')| \geqslant \varepsilon$. Then in particular, writing $\delta = \dfrac{1}{n}$ we conclude that two sequences $\{x_n\}$ and $\{x'_n\}$ exist satisfying the following inequalities:

(7) $\qquad a \leqslant x_n \leqslant b, \qquad a \leqslant x'_n \leqslant b. \qquad |x_n - x'_n| < \dfrac{1}{n}$

$$\text{and} \qquad |f(x_n) - f(x'_n)| \geqslant \varepsilon.$$

The sequence $\{x_n\}$ being bounded, it contains according to the Bolzano-Weierstrass theorem (§ 2.6, 2) a convergent subsequence. Let us write $\lim_{n=\infty} x_{m_n} = c$. It follows from the first of the inequalities (7) that $a \leqslant c \leqslant b$. Hence the function f is continuous at the point c and so $\lim_{n=\infty} f(x_{m_n}) = f(c)$. But the third of the inequalities (7) implies also $\lim_{n=\infty} x'_{m_n} = c$, because $\lim_{n=\infty} x'_{m_n} = \lim_{n=\infty} x_{m_n}$. The continuity of the function f gives $\lim_{n=\infty} f(x'_{m_n}) = f(c)$ as before. Thus $\lim_{n=\infty} [f(x_{m_n}) - f(x'_{m_n})] = 0$. But the last equation is contradictory to the last of the inequalities (7).

Hence, the theorem is proved.

Remark. The assumption that the considered interval is closed is essential; e.g. the function $\dfrac{1}{x}$ is continuous at any point of the open interval $0 < x < 1$ but it is not uniformly continuous in this interval.

2. WEIERSTRASS THEOREM. *A function f continuous in a closed interval $a \leqslant x \leqslant b$ is bounded and attains its upper and lower bounds M and m, i.e. there exist in this interval two points c and d such that $f(c) = M$ and $f(d) = m$.*

First of all we shall prove the boundedness of the function f, i.e. that a number A exists such that the inequality $A > |f(x)|$ holds for any x belonging to the interval ab. Now, by substitution $\varepsilon = 1$, Theorem 1 implies the existence of a $\delta > 0$ such that if two points x and x' belong to an interval (contained in the interval ab) of the length $< \delta$, then $|f(x) - f(x')| < 1$. Let us choose a number n in such way that $\dfrac{b-a}{n} < \delta$; dividing the interval ab in n equal intervals, the length of everyone of them will be less than δ. Thus, denoting the ends of these intervals by $a_0, a_1, \ldots, a_n$ ($a_0 = a$, $a_n = b$), successively, we have

$$|f(x) - f(a_1)| < 1 \quad \text{for} \quad a_0 \leqslant x \leqslant a_1,$$
$$\text{whence} \quad |f(x)| < 1 + |f(a_1)|,$$

and in general:
$$|f(x)-f(a_k)| < 1 \quad \text{for} \quad a_{k-1} \leqslant x \leqslant a_k,$$
whence
$$|f(x)| < 1 + |f(a_k)|.$$

Thus, denoting by A the greatest of the numbers $1+|f(a_k)|$, where k assumes the values $1, 2, \ldots, n$, we have $|f(x)| < A$ for every x belonging to the interval ab.

So we have proved that the function f is bounded, i.e. that the set of values of this function is bounded. Hence, the upper bound M and the lower bound m of this set exist (cf. § 1.7). We shall prove by *reductio ad absurdum* that M is one of the values of the function f (the proof for m is completely analogous).

Let us suppose that it is not true, i. e. that we have $M-f(x) \neq 0$ for any x. Then the function $g(x) = \dfrac{1}{M-f(x)}$ is defined for any x in the interval $a \leqslant x \leqslant b$ and is continuous in this interval. Hence it is a bounded function, as we have just proved. Then a number N exists such that $g(x) < N$, i.e. $M-f(x) > \dfrac{1}{N}$, that is $f(x) < M - \dfrac{1}{N}$. But the last inequality proves the existence of a number less than M which is greater than all numbers $f(x)$, where x takes values from the interval $a \leqslant x \leqslant b$. But this is contradictory to the assumption that M is the upper bound of the function f in this interval.

Thus, supposing the theorem to be wrong we have got a contradiction.

3. DARBOUX [1] PROPERTY. *A function continuous in a closed interval $a \leqslant x \leqslant b$ assumes all values between any two values of this function.* This means that, if y is a number between $f(a)$ and $f(b)$ (i.e. either $f(a) < y < f(b)$ or $f(b) < y < f(a)$), then there exists a c in the interval ab such that $f(c) = y$.

---

[1] Gaston Darboux (1842-1917), a famous French geometer.

Let us assume that $f(a) < y < f(b)$ (in the case of the other eventuality, the arguments are completely similar). Supposing the theorem to be false, i.e. supposing $y-f(x) \neq 0$ for any x belonging to the interval ab, the function $h(x) = \dfrac{1}{|y-f(x)|}$ would be bounded by Theorem 2. Let $M > h(x)$, i.e. let

(8) $$|y-f(x)| > \frac{1}{M}.$$

Substituting $\varepsilon = \dfrac{1}{M}$ in Theorem 1 we conclude that a number $\delta > 0$ exists such that if two points x and x' belong to an interval of the length $< \delta$, then

$$|f(x)-f(x')| < \frac{1}{M}.$$

Let n indicate a positive integer such that $\dfrac{b-a}{n} < \delta$. Dividing the segment ab into n equal segments and denoting by $a_0, a_1, \ldots, a_n$ the successive ends of these segments, as in the proof of the previous theorem, we conclude that

(9) $\quad |f(a_k)-f(a_{k-1})| < \dfrac{1}{M} \quad$ for $\quad k = 1, 2, \ldots, n$.

Since $f(a_0) < y < f(a_n)$, so there exists among the numbers $1, 2, \ldots, n$ the least number m satisfying the condition $y < f(a_m)$. We have $m > 0$ and $f(a_{m-1}) < y < f(a_m)$, whence

$$0 < y - f(a_{m-1}) < f(a_m) - f(a_{m-1}) < \frac{1}{M}$$

by (9), but this contradicts the inequality (8).

Theorems 2 and 3 lead to the following conclusion:

4. *A function continuous in a closed interval $a \leqslant x \leqslant b$ assumes all values between its lower bound m and upper*

bound M, including the bounds. In other words, the set of values of the function is the closed interval $m \leqslant y \leqslant M$.

Remarks. (α) The property of continuous functions expressed in Theorem 3 (the so-called Darboux property) is especially intuitive in its geometrical interpretation. However, this property does not characterize continuous functions, e.g. the function (5) possesses this property although it is not continuous.

(β) Theorem 3 implies immediately the existence of $\sqrt[n]{x}$ for any real x, if n is odd and for any $x \geqslant 0$, if n is even. Indeed, if n is odd, then $\lim\limits_{x=-\infty} x^n = -\infty$ and $\lim\limits_{x=\infty} x^n = \infty$ and so the function x^n assumes all real values. Thus the function inverse with regard to x^n, i.e. $\sqrt[n]{x}$ is defined for every real x. Similarly, if n is even, then the function x^n, $x \geqslant 0$, assumes all values from 0 to ∞; hence $\sqrt[n]{x}$ is defined for all these values.

Analogously, it follows from the formulae $\lim\limits_{x=-\infty} e^x = 0$ and $\lim\limits_{x=\infty} e^x = \infty$ that the function e^x assumes all positive values; hence $\log x$ is defined for any positive x. Finally, $\arctan x$ is defined for every real x.

Let us add that the formulae § 4.4, (3) imply immediately the following formulae connecting logarithms, for positive a, x and y ($a \neq 1$):

(10) $\log_a xy = \log_a x + \log_a y$, $\log_a (x^y) = y \log_a x$,

(11) $a^x = e^{x \log a}$, $\log_a x = \dfrac{\log x}{\log a}$.

For example to prove the first of the formulae (10), let us substitute $\log_a x = t$, $\log_a y = z$, i.e. $x = a^t$, $y = a^z$. By § 4.4, (3) we get

$xy = a^t \cdot a^z = a^{t+z}$, whence $\log_a xy = t + z = \log_a x + \log_a y$.

The second of the formulae (11) is obtained from the second of the formulae (10):

$\log a^t = t \log a$,   i.e. $\log x = \log_a x \log a$.

5.5. Continuity of inverse functions

In § 4.3 we have seen that a one-to-one function $y = f(x)$ defines an inverse function $x = g(y)$. Now, we shall prove that *a function inverse with regard to a continuous function is also continuous*. Speaking more exactly:

1. *If a function $y = f(x)$ is one-to-one and continuous in an interval $a \leqslant x \leqslant b$, then the inverse function $x = g(y)$ is continuous in the interval $m \leqslant y \leqslant M$, where m and M are the lower bound and the upper bound of the function f, respectively.*

Let $m \leqslant c \leqslant M$. By Theorem 4 of § 5.4, the function g is defined at the point c. Let $c = \lim_{n=\infty} y_n$, where y_n belongs to the interval mM, i.e. y_n is of the form $y_n = f(x_n)$. It has to be proved that $\lim_{n=\infty} g(y_n) = g(c)$.

Let $c = f(d)$. Thus we have to prove that the condition $\lim_{n=\infty} f(x_n) = f(d)$ implies $\lim_{n=\infty} x_n = d$ (because $g(y_n) = x_n$, $g(c) = d$). Since the sequence $\{x_n\}$ is bounded, being contained in the interval ab, so the last inequality will be proved, if we show that every convergent subsequence of the sequence $\{x_n\}$ is convergent to the limit d (cf. § 2.6, Theorem 4). Thus, let $\lim_{n=\infty} x_{k_n} = d'$. We shall prove that $d' = d$. The continuity of the function f implies $\lim_{n=\infty} f(x_{k_n}) = f(d')$ and since $\lim_{n=\infty} f(x_{k_n}) = \lim_{n=\infty} f(x_n) = f(d)$, we get $f(d') = f(d)$. The function f is one-to-one, hence we conclude that $d' = d$.

Thus, the theorem is proved.

APPLICATIONS. (α) $\log_a x$ *is a continuous function* $(1 \neq a > 0)$, for it is inverse to the continuous function a^x.

(β) *The cyclometric functions* $\arcsin x$, $\arccos x$, $\arctan x$ *etc. are continuous* as inverse functions with regard to the trigonometric functions (which are continuous).

5. CONTINUOUS FUNCTIONS

(γ) *Let $\{a_n\}$ be a sequence with positive terms. Then the convergence of the infinite product $\prod_{n=1}^{\infty} a_n$ implies the convergence of the infinite series $\sum_{n=1}^{\infty} \log a_n$. Moreover, there holds the relation*

(12) $\quad \log \prod_{n=1}^{\infty} a_n = \sum_{n=1}^{\infty} \log a_n, \quad i.e. \quad \prod_{n=1}^{\infty} a_n = e^{\sum_{n=1}^{\infty} \log a_n}$

Similarly, *assuming the convergence of the series $\sum_{n=1}^{\infty} \log a_n$, the product $\prod_{n=1}^{\infty} a_n$ is convergent, too, and the relation* (12) *takes also place.*

Let $p_n = a_1 \cdot a_2 \cdot \ldots \cdot a_n$. The assumption of convergence of the sequence $\{p_n\}$ and the continuity of the logarithm imply (cf. (2)):

$$\log \lim_{n=\infty} p_n = \lim_{n=\infty} \log p_n$$

and

$$\lim_{n=\infty} \log p_n = \lim_{n=\infty} \log(a_1 \cdot \ldots \cdot a_n) = \lim_{n=\infty} [\log a_1 + \ldots + \log a_n]$$
$$= \sum_{n=1}^{\infty} \log a_n.$$

Similarly, writing $s_n = \log a_1 + \ldots + \log a_n$ and assuming the convergence of the sequence $\{s_n\}$, we have by the continuity of the exponential function:

$$e^{\lim_{n=\infty} s_n} = \lim_{n=\infty} e^{s_n},$$

but

$$e^{s_n} = e^{\log a_1 + \ldots + \log a_n} = e^{\log a_1} \cdot \ldots \cdot e^{\log a_n} = p_n,$$

whence

$$\lim_{n=\infty} e^{s_n} = \lim_{n=\infty} p_n.$$

The formula (12) being, as easily seen, a generalization of the first of the formulae (10) to the infinite operations enables us to transfer many theorems concerning infinite series to the case of infinite products.

2. *Every one-to-one function f continuous in an interval $a \leqslant x \leqslant b$ is strictly monotone (i.e. either increasing or decreasing).*

Our assumption implies $f(a) \neq f(b)$ and we may further assume that $f(a) < f(b)$ (if $f(a) > f(b)$, then analogous arguments are applicable). We shall prove the function f to be increasing in $a \leqslant x \leqslant b$. Let $x < x'$. One has to prove that $f(x) < f(x')$.

First of all let us note that the conditions $a \leqslant x \leqslant b$ and $f(a) < f(b)$ imply $f(a) \leqslant f(x) \leqslant f(b)$. Indeed, otherwise we should have either $f(x) < f(a)$ or $f(x) > f(b)$. In the first case, the double inequality $f(x) < f(a) < f(b)$ would imply (by Theorem 3) the existence of a point x'' in the interval xb such that $f(x'') = f(a)$. But this contradicts the assumption that f is one-to-one (because $x'' \neq a$). Similarly, in the other case a point x''' would exist in the interval ax such that $f(x''') = f(b)$, in contradiction to the inequality $x''' \neq b$.

So we have proved that $f(a) \leqslant f(x) \leqslant f(b)$. At the same time we conclude that the conditions $x \leqslant x' \leqslant b$ and $f(x) < f(b)$ imply $f(x) \leqslant f(x') \leqslant f(b)$.

Thus, $f(x) < f(x')$, as we had to prove.

Exercises on § 5

1. Let $f(x)$ denote a function defined as follows: if x is an irrational number, then $f(x) = 0$, and if x is a rational number $x = \dfrac{p}{q}$ (the fraction $\dfrac{p}{q}$ being irreducible), then $f(x) = \dfrac{1}{q}$. Prove that this function is continuous at irrational points and discontinuous at rational points.

5. CONTINUOUS FUNCTIONS

2. Prove that every function $f(x)$, $a \leqslant x \leqslant b$, one-to-one and possessing the Darboux property (§ 5.4, Remark (α)) is continuous.

3. Prove that every equation of an odd degree n with real coefficients, i.e. equation of the form $x^n + a_{n-1}x^{n-1} + \ldots + a_1 x + a_0 = 0$ has at least one real root (cf. Exercise 2, § 4).

4. Prove that a continuous function satisfying the condition
$$f(x+y) = f(x) + f(y)$$
for any x and y is of the form $f(x) = ax$.
(Prove this equation for rational x, taking $a = f(1)$)

5. Find whether the function $\sin\dfrac{1}{x}$ is uniformly continuous in the interval $0 < x < 1$, and the function $\sqrt{x}$ in the interval $0 \leqslant x < \infty$.

6. Let $f(x)$ and $g(x)$ be two continuous functions in an interval $a \leqslant x \leqslant b$. Let $h(x)$ denote the greater of the two values $f(x)$ and $g(x)$ (eventually their common value, if they are equal). Prove that $h(x)$ is continuous.

7. Given a sequence $a_1, a_2, \ldots$, choose a function discontinuous at the points of this sequence and continuous at all other points.

8. A function f is said to satisfy the *Lipschitz condition*, if a constant C exists such that the inequality
$$|f(x) - f(x')| \leqslant C|x - x'|$$
holds for any pair x and x'.

Characterize the Lipschitz condition geometrically.

Prove that a function satisfying the Lipschitz condition (in an interval which may be closed or open, bounded or unbounded) is uniformly continuous.

Show by an example that the converse theorem does not hold.

§ 6. SEQUENCES AND SERIES OF FUNCTIONS

6.1. Uniform convergence

If to any positive integer n there corresponds a function f_n, then a sequence of functions $f_1, f_2, ..., f_n, ...$ is defined. Let us assume all these functions to be defined in the same set of arguments A (e.g. in an interval $a \leqslant x \leqslant b$). If the sequence $f_1(x), f_2(x), ...$ is convergent for every x (belonging to A) to a limit $f(x)$, then we say that *the sequence of functions $\{f_n\}$ is convergent to the function f*. This means by the definition of the limit (§ 2.2) that to every x and to any $\varepsilon > 0$ a number k exists such that for $n > k$ the inequality

(1) $$|f_n(x) - f(x)| < \varepsilon$$

holds.

Thus, given $\varepsilon > 0$, the number k is chosen suitably to any x, separately. However, if k can be chosen *independently* of x, then the convergence of our sequence is said to be *uniform*. Thus, *a sequence $\{f_n\}$ is uniformly convergent to a function f* (in a set A), if to any $\varepsilon > 0$ a number k exists such that for each $n > k$ and for every x (belonging to A) the inequality (1) holds.

Geometrically, the uniform convergence means that surrounding the limit-curve $y = f(x)$ with a strip defined by two curves parallel to it (given by the equations $y = f(x) \pm \varepsilon$), all curves $y = f_n(x)$ with sufficiently large n are contained in this strip.

EXAMPLES. Let $f_n(x) = x^n$. We have

$$\lim_{n=\infty} f_n(x) = 0 \quad \text{for} \quad 0 \leqslant x < 1$$
$$\text{and} \quad \lim_{n=\infty} f_n(x) = 1 \quad \text{for} \quad x = 1.$$

This sequence is uniformly convergent in the interval $0 \leqslant x \leqslant \tfrac{1}{2}$.

For, given $\varepsilon > 0$, let us choose a number k in such way that $\frac{1}{2^k} < \varepsilon$. Then assuming $x \leqslant \frac{1}{2}$ we have $x^n \leqslant \left(\frac{1}{2}\right)^n$ $< \frac{1}{2^k} < \varepsilon$ for $n > k$ and the inequality (1) is satisfied by each $n > k$ and by any x belonging to the interval $0 \leqslant x \leqslant \frac{1}{2}$. But this means that the convergence of the sequence $\{f_n\}$ in this interval is uniform. Yet the convergence in the interval $0 \leqslant x \leqslant 1$ is not uniform. This is easily seen substituting e.g. $\varepsilon = \frac{1}{2}$; moreover, it follows immediately from the discontinuity of the limit-function and from the following theorem:

1. *A limit of a uniformly convergent sequence of continuous functions is continuous.*

Let the sequence $\{f_n\}$ be uniformly convergent to f (in a set A) and let $\varepsilon > 0$. Then there is a k such that for each $n \geqslant k$, the inequality

(2) $\qquad |f_n(x) - f(x)| < \frac{1}{3}\varepsilon$

is satisfied for every x (belonging to A). Let a be an arbitrary point belonging to A. We have to prove that the function f is continuous at this point. Thus, given ε, a number $\delta > 0$ has to be chosen in such a way that the inequality $|x-a| < \delta$ would imply

(3) $\qquad |f(x) - f(a)| < \varepsilon$

(for x belonging to A).

Now, f_k being continuous at the point a, a number $\delta > 0$ exists such that the inequality $|x-a| < \delta$ implies

(4) $\qquad |f_k(x) - f_k(a)| < \frac{1}{3}\varepsilon$

The inequality (2) being satisfied for $n = k$ and $x = a$, we find that

(5) $\qquad |f_k(a) - f(a)| < \frac{1}{3}\varepsilon$.

Replacing n in the inequality (2) by k and adding to the inequality obtained in this way the inequalities (4) and (5), we get the inequality (3), as was required.

Remark. As we have seen, the function $f(x) = \lim\limits_{n=\infty} x^n$ (cf. Fig. 9) is discontinuous in the interval $0 \leqslant x \leqslant 1$. Thus, the limit of a non-uniformly convergent sequence of continuous functions may be discontinuous. Theorem 1 and the above example illustrate the meaning of the notion of uniform convergence.

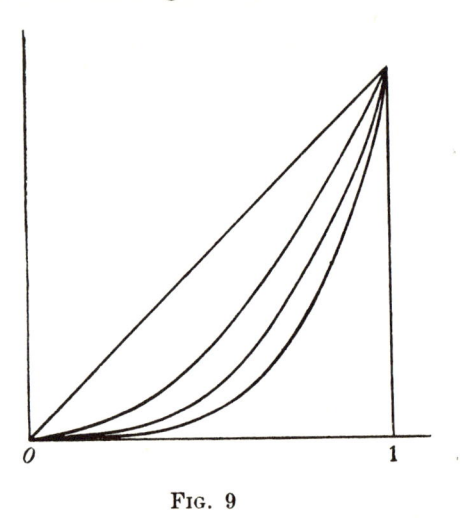

Fig. 9

Similarly as in the case of the uniform continuity of a function, the expression "uniform" shows that the number k chosen to a given ε is independent of x.

The following theorem corresponds to the Cauchy condition for the convergence of a sequence:

2. *A necessary and sufficient condition for the uniform convergence of a sequence of functions $\{f_n\}$ is that to any $\varepsilon > 0$ a number r exists such that the inequality*

(6) $$|f_n(x) - f_r(x)| < \varepsilon$$

holds for each $n > r$.

First let us assume the sequence to be uniformly convergent to a function f. Then, given a number $\varepsilon > 0$,

there exists an r such that for each $n \geqslant r$, we have
$$|f_n(x)-f(x)| < \tfrac{1}{2}\varepsilon, \quad \text{whence} \quad |f_r(x)-f(x)| < \tfrac{1}{2}\varepsilon.$$

Adding both these inequalities we obtain the inequality (6).

Now, let us assume a sequence $\{f_n\}$ to satisfy the condition formulated in the theorem. Then the sequence $f_1(x), f_2(x), \ldots$ satisfies for a fixed x the Cauchy condition; thence it is convergent. Speaking otherwise, the function $f(x) = \lim\limits_{n=\infty} f_n(x)$ exists. It has to be proved that the convergence is uniform. Let an $\varepsilon > 0$ be given. According to the assumption, an r exists such that the inequality

(7) $$|f_n(x)-f_r(x)| < \tfrac{1}{2}\varepsilon$$

holds for each $n > r$.

We shall prove that for $n > r$, the inequality (1) holds, too (with the sign $\leqslant$ instead of $<$).

Replacing the variable n by m in formula (7) and adding to the inequality (7) the obtained inequality $|f_m(x)-f_r(x)| < \tfrac{1}{2}\varepsilon$, we get for $m > r$,
$$|f_n(x)-f_m(x)| < \varepsilon, \quad \text{whence} \quad \lim_{m=\infty}|f_n(x)-f_m(x)| \leqslant \varepsilon,$$

i.e.
$$|\lim_{m=\infty}[f_n(x)-f_m(x)]| \leqslant \varepsilon, \quad \text{that is} \quad |f_n(x)-\lim_{m=\infty}f_m(x)| \leqslant \varepsilon$$
and so
$$|f_n(x)-f(x)| \leqslant \varepsilon.$$

6.2. Uniformly convergent series

Let a sequence of functions $f_1, f_2, \ldots, f_n, \ldots$ defined on the same set of arguments be given.

The series $f_1(x)+f_2(x)+\ldots+f_n(x)+\ldots$ is called *uniformly convergent*, if the sequence of its partial sums
$$s_n(x) = f_1(x)+f_2(x)+\ldots+f_n(x)$$
is uniformly convergent.

If the functions $f_1, f_2, \ldots$ are continuous, then the partial sums s_n are obviously also continuous. By the theorems of § 6.1, we therefore conclude that

1. *The sum of a uniformly convergent sequence of continuous functions is a continuous function.*

2. *A necessary and sufficient condition for the uniform convergence of a series $f_1(x) + f_2(x) + \ldots$ is that to any number $\varepsilon > 0$ a number k exists such that the inequality*

$$(8) \qquad |f_k(x) + f_{k+1}(x) + \ldots + f_n(x)| < \varepsilon$$

holds for each $n > k$ and for any x.

From Theorem 2 we may deduce the following theorem which makes it possible to infer the uniform convergence of a series $\sum_{n=1}^{\infty} f_n(x)$ by comparison of the terms of this series with the terms of a series $\sum_{n=1}^{\infty} u_n$.

3. *If a series $\sum_{n=1}^{\infty} u_n$ is convergent and if the inequality $|f_n(x)| \leqslant u_n$ is satisfied for any x, then the series $\sum_{n=1}^{\infty} f_n(x)$ converges uniformly and absolutely.*

Indeed, since the series $u_1 + u_2 + \ldots$ is convergent, therefore to any given $\varepsilon > 0$ a number k exists such that the inequality $u_k + u_{k+1} + \ldots + u_n < \varepsilon$ is satisfied for each $n > k$. Hence by the assumption,

$$|f_k(x) + \ldots + f_n(x)| \leqslant |f_k(x)| + \ldots + |f_n(x)| \leqslant u_k + \ldots + u_n < \varepsilon;$$

thus the formula (8) holds and, by Theorem 2, this implies the uniform convergence of the series $\sum_{n=1}^{\infty} f_n(x)$.

The second part of the above double inequality establishes at once the absolute convergence of the series.

EXAMPLE. The series $\sum_{n=0}^{\infty} \dfrac{x^n}{n!}$ is uniformly convergent in every interval $-a \leqslant x \leqslant a$, since $\left|\dfrac{x^n}{n!}\right| \leqslant \dfrac{a^n}{n!}$ and the series $\sum_{n=0}^{\infty} \frac{a^n}{n!}$ is convergent for any a (§ 3.7).

We shall encounter many more examples of uniformly convergent series in § 6.3 below.

6.3. Power series

A series of the form

$$(9) \quad S(x) = a_0 + a_1 x + a_2 x^2 + \ldots + a_n x^n + \ldots = \sum_{n=0}^{\infty} a_n x^n \quad (^1)$$

is called a *power series*.

Thus we see that a power series is a natural generalization of the notion of a polynomial of degree n. The importance of power series follows also from the fact that the representation of a function by means of a power series is very advantageous from the point of view of practical calculations.

Let us consider the following three examples of power series:

(a) $1 + \frac{x}{1!} + \frac{x^2}{2!} + \frac{x^3}{3!} + \ldots,$ (b) $1 + x + x^2 + x^3 + \ldots,$

(c) $1 + 1! x + 2! x^2 + 3! x^3 + \ldots$

As well known, the first of these series is convergent for all x. The second one is convergent for $|x| < 1$ and its sum $= \frac{1}{1-x}$; for other x the series is divergent. Finally, the third series is divergent for every $x \neq 0$.

In order to formulate a general theorem on the convergence of power series, the notion of the "radius of convergence" of a power series will be introduced. Namely, the *radius of convergence* of a power series $S(x)$ is defined

---

[1] Strictly speaking, for $x = 0$ the symbol $\sum_{n=0}^{\infty} a_n x^n$ (which is in this case indefinite) has to be replaced by a_0 (cf. the footnote to the formula (6), § 1).

as the upper bound of the set of absolute values of x for which this series is convergent. In particular, if the above set is unbounded, then the radius of convergence is assumed to be equal to ∞.

For example the radii of convergence of the series (a), (b) and (c) are ∞, 1 and 0, respectively.

The interval $-r < x < r$, where r denotes the radius of convergence of a given power series, is called its *interval of convergence*. In the case $r = \infty$, the interval of convergence is equal to the set of all real numbers. This terminology is connected with the following theorem:

1. *In any closed interval lying inside the interval of convergence the power series is convergent uniformly and absolutely.*

It suffices to prove our theorem for intervals of the form $-c \leqslant x \leqslant c$, where $c < r$ (because every closed interval lying inside the interval of convergence is contained in such an interval with the centre 0). Since r, being the radius of convergence, is the upper bound of the set of absolute values of the convergence-points of the series $S(x)$, so a point $b > c$ exists such that either $S(b)$ or $S(-b)$ is convergent. It follows from the convergence of the series $S(\pm b)$ that its terms constitute a bounded sequence. Thus, let $M > |a_n b^n|$ for each n. Hence we conclude that for $|x| \leqslant c$, we have

$$|a_n x^n| = |a_n b^n| \cdot \left|\frac{x}{b}\right|^n \leqslant M \left(\frac{c}{b}\right)^n.$$

Since the series $\sum\limits_{n=0}^{\infty} \left(\frac{c}{b}\right)^n$, being a geometric series with quotient <1, is convergent, Theorem 3 of § 6.2 implies the uniform and absolute convergence of the series $S(x)$ in the interval $-c \leqslant x \leqslant c$.

Remark. We conclude from the above theorem that the series is convergent inside its interval of convergence and divergent outside this interval (the last follows from the definition of the radius of convergence). Yet at the ends of the interval of convergence the series may be convergent as well as divergent. The following examples give evidence here: the series (b) is divergent at the ends of the interval of convergence, namely, $S(1) = \infty$ and the series $S(-1) = 1-1+1-1+...$ has neither a finite nor an infinite limit; the series

$$\frac{x}{1} - \frac{x^2}{2} + \frac{x^3}{3} - \frac{x^4}{4} + ...$$

(representing $\log(1+x)$) is convergent for $x = 1$ and divergent to $-\infty$ for $x = -1$.

The previous theorem together with Theorem 1 of § 6.2 imply:

2. *A power series $S(x)$ is a continuous function in the open interval $-r < x < r$, i.e. inside the interval of convergence.*

Indeed, to any x lying between $-r$ and r there may be chosen a closed interval with the centre x, lying inside the interval of convergence. Since the power series is uniformly convergent in this closed interval, it follows from Theorem 1 of § 6.2, that it represents a continuous function.

Applying Abel's theorem from the theory of infinite series (§ 3.3, Theorem 2) we shall complete this theorem with respect to the continuity of the power series at the ends of the interval of convergence.

3. ABEL THEOREM. *A power series convergent at one of the ends of its interval of convergence constitutes a (one-side) continuous function at this point.*

Speaking more strictly, if the series $S(r)$ (or $S(-r)$) is convergent, then the series $S(x)$ is uniformly convergent in the interval $0 \leqslant x \leqslant r$ (or in the interval $-r \leqslant x \leqslant 0$).

Let us assume the series $S(r)$ to be convergent and let us denote by R_n the n-th remainder of this series, i.e.

$$R_n = a_{n+1}r^{n+1} + a_{n+2}r^{n+2} + ...$$

Then we have $\lim_{n=\infty} R_n = 0$. Let an $\varepsilon > 0$ be given. Thus a number k exists such that we have

$$|a_{n+1}r^{n+1} + \ldots + a_i r^i| < \varepsilon$$

for $i > n > k$.

Let us denote by $R_n(x)$ the n-th remainder of the series $S(x)$:

$$R_n(x) = a_{n+1}x^{n+1} + a_{n+2}x^{n+2} + \ldots$$

Hence, in particular, $R_n(r) = R_n$. Let us note that

$$R_n(x) = a_{n+1}r^{n+1}\left(\frac{x}{r}\right)^{n+1} + a_{n+2}r^{n+2}\left(\frac{x}{r}\right)^{n+2} + \ldots$$

and let us apply Abel's theorem (§ 3.3, 2 and Remark (α)) after replacing the sequence $\{a_n\}$ by the sequence $\left(\frac{x}{r}\right)^{n+1}$, $\left(\frac{x}{r}\right)^{n+2}, \ldots$ and the series $b_1 + b_2 + \ldots$ by the series R_n.

Since this series is bounded by the number ε and since

$$1 > \left(\frac{x}{r}\right)^{n+1} > \left(\frac{x}{r}\right)^{n+2} > \ldots \quad \text{and} \quad \lim_{m=\infty}\left(\frac{x}{r}\right)^m = 0 \quad \text{for} \quad 0 \leqslant x < r,$$

we conclude that $|R_n(x)| < 2\varepsilon$. This inequality is satisfied also for $x = r$. Thus we have shown that if $n > k$, then the inequality

$$\left|\sum_{m=0}^{\infty} a_m x^m - \sum_{m=0}^{n} a_m x^m\right| < 2\varepsilon$$

is satisfied for any x such that $0 \leqslant x \leqslant r$. This means that the series $S(x)$ is uniformly convergent in this interval.

In the case when the series $S(-r)$ is convergent the arguments are analogous.

Theorems 2 and 3 may be expressed also in the following way: *every point of convergence of a power series is a point of continuity.*

Sometimes the d'Alembert and Cauchy criteria enable us to calculate the radius of convergence easily. Namely:

4. If $\lim_{n=\infty} \left|\frac{a_{n+1}}{a_n}\right| = g$, then the radius of convergence of the power series $S(x)$ is equal to $r = 1/g$ (moreover, if $g = 0$, then $r = \infty$ and if $g = \infty$, then $r = 0$).

Analogously, if $\lim_{n=\infty} \sqrt[n]{|a_n|}$ exists, then $r = \dfrac{1}{\lim\limits_{n=\infty} \sqrt[n]{|a_n|}}$.

Indeed, let $x > 0$. We obtain

(10) $$\lim_{n=\infty} \left|\frac{a_{n+1} x^{n+1}}{a_n x^n}\right| = x \cdot \lim_{n=\infty} \left|\frac{a_{n+1}}{a_n}\right| = xg.$$

Hence if $x < 1/g$, then the considered limit is < 1 and so the series $S(x)$ is convergent (absolutely), whence $r \geqslant 1/g$. Supposing $r > 1/g$, the series $S(x)$ would be absolutely convergent for $1/g < x < r$. But this is impossible, since then the limit considered in the formula (10) would be > 1.

The above arguments concern also the cases $g = 0$ or $g = \infty$.

The proof in the case of the Cauchy criterion is similar.

6.4. Approximation of continuous functions by polygonal functions

A function $f(x)$, $a \leqslant x \leqslant b$, is called *polygonal*, if the interval ab may be divided into a finite number of smaller intervals in such a way that the function is linear in everyone of these intervals, separately. This means that there exist a system of $n+1$ points

(11) $\quad a_0 < a_1 < \ldots < a_n$, where $a_0 = a$, $a_n = b$,

and two systems of real numbers $c_1, c_2, \ldots, c_n$ and $d_1, d_2, \ldots, d_n$ so that

(12) $\quad f(x) = c_k x + d_k \quad$ for $\quad a_{k-1} \leqslant x \leqslant a_k$
$$(k = 1, 2, \ldots, n).$$

For example the function $f(x) = |x|$ is polygonal in the interval $-1 \leqslant x \leqslant 1$, for $f(x) = -x$ in the interval $-1, 0$, and $f(x) = x$ in the interval $0, 1$.

The graph of a polygonal function is a polygonal line. Conversely, if a polygonal line is the graph of a function, then this function is a polygonal one.

Sometimes we are dealing with approximation of curves by polygonal lines (e.g. in the elementary geometry, calculating the length of the arc of a circle, we approximate this arc by inscribed and circumscribed polygonal lines).

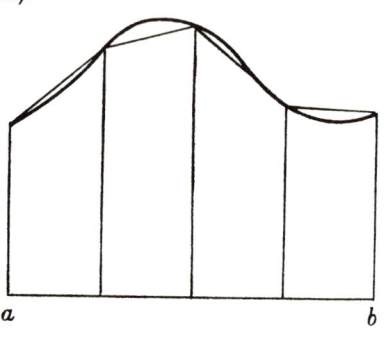

Fig. 10

The following theorem expresses analytically the approximation of curves given by continuous functions by polygonal lines (given by polygonal functions):

THEOREM. *Every continuous function f in an interval $a \leqslant x \leqslant b$ is the limit of a uniformly convergent sequence of polygonal functions.*

For each n, we consider the partition of the segment ab into n equal segments: $aa_1, a_1a_2, ..., a_{n-1}b$ (thus we have in general $a_k = a + k\dfrac{b-a}{n}$). We draw a polygonal line through the points $[a, f(a)], [a_1, f(a_1)], ..., [a_n, f(a_n)]$. This polygonal line is the graph of a certain polygonal function f_n. We shall prove that the sequence of the functions f_n is uniformly convergent to the function f.

According to the uniform continuity of the function f in the interval ab, to any given $\varepsilon > 0$ a number $\delta > 0$ may be chosen so that the inequality $|x-x'| < \delta$ implies $|f(x)-f(x')| < \varepsilon$. Let $m > \dfrac{b-a}{\delta}$. We shall prove that for $n > m$ we have the inequality

(13) $\qquad |f_n(x)-f(x)| < 2\varepsilon$

for every x belonging to the interval ab.

Indeed, let x be given and let $n > m$. Moreover, let the point x belong to the interval $a_{k-1}a_k$ (of the n-th partition). Since the length of this interval is $< \delta$, so

(14) $\quad |f(a_{k-1})-f(a_k)| < \varepsilon \quad$ and $\quad |f(x)-f(a_k)| < \varepsilon$.

Let us suppose that $f(a_{k-1}) \leqslant f(a_k)$ (if $f(a_{k-1}) \geqslant f(a_k)$, then the arguments are analogous). Then the function f_n is non-decreasing in $a_{k-1}a_k$, since it is linear. Thus

(15) $\qquad f_n(a_{k-1}) \leqslant f_n(x) \leqslant f_n(a_k)$.

Since, according to the definition of the function f_n, we have $f_n(a_{k-1}) = f(a_{k-1})$ and $f_n(a_k) = f(a_k)$, so $f(a_{k-1}) \leqslant f_n(x) \leqslant f(a_k)$ and this gives, according to the first of the inequalities (14), $|f_n(x)-f(a_k)| < \varepsilon$. Comparing this inequality with the second of the inequalities (14) we obtain the formula (13).

We shall now cite without proof the following Weierstrass theorem, analogous to the previous one:

Every function continuous in an interval $a \leqslant x \leqslant b$ is the limit of a uniformly convergent sequence of polynomials.

6.5*. The symbolism of mathematical logic

In order to understand better the notions and the theorems of mathematical analysis, a knowledge of the notation of definitions and theorems by means of some symbols used in mathematical logic is required.

We shall give here the principal notions and symbols employed in mathematical analysis.

Given two sentences α and β, we denote by $\alpha \vee \beta$ the sentence "α or β" (the sum, i.e. the alternative of the sentences α and β); $\alpha \wedge \beta$ will denote the sentence "α and β" (the product, i.e. the conjunction of the sentences α and β).

Every sentence has one of the two values: truth (indicated by the figure 1) and falsehood (indicated by 0). A logical sum $\alpha \vee \beta$ is a true sentence, if and only if at least one of its terms is a true sentence. Similarly, a logical product $\alpha \wedge \beta$ is a true sentence, if and only if both its factors are true sentences.

It is easy to show that logical addition and multiplication possess the properties of the commutativity and the associativity:

$$\alpha \vee \beta = \beta \vee \alpha, \quad \alpha \wedge \beta = \beta \wedge \alpha, \quad \alpha \vee (\beta \vee \gamma) = (\alpha \vee \beta) \vee \gamma,$$
$$\alpha \wedge (\beta \wedge \gamma) = (\alpha \wedge \beta) \wedge \gamma.$$

The laws of distributivity also hold:

$$\alpha \wedge (\beta \vee \gamma) = (\alpha \wedge \beta) \vee (\alpha \wedge \gamma)$$
$$\text{and} \quad \alpha \vee (\beta \wedge \gamma) = (\alpha \vee \beta) \wedge (\alpha \vee \gamma).$$

Besides the above two operations, the negation of the sentence α, indicated by $\sim \alpha$, is also considered. It is clear that the sentence $\sim \alpha$ is true, if the sentence α is false, and is false, if the sentence α is true. Moreover, it is immediately seen that

(16) $$\sim(\sim \alpha) = \alpha$$

(two negations are reducing themselves).

The following two theorems (so-called *de Morgan laws*) may also be easily proved:

(17) $\quad \sim(\alpha \vee \beta) = (\sim \alpha) \wedge (\sim \beta)$
$$\text{and} \quad \sim(\alpha \wedge \beta) = (\sim \alpha) \vee (\sim \beta).$$

The first of these two theorems states that the sentence that one of the two sentences α and β is true, is a false sentence if and only if both sentences $\sim \alpha$ and $\sim \beta$ are true, i.e. if and only if both sentences α and β are false.

6. SEQUENCES AND SERIES OF FUNCTIONS 131

The second theorem states that "it is not true that both sentences α and β are true, simultaneously", if and only if at least one of these sentences is false.

Besides the operations of addition, multiplication and negation, the relation of implication is considered in the logic of sentences. Namely, we write $\alpha \to \beta$ in order to denote that the sentence α implies the sentence β, i.e. that if α is true, then β is true, too; in other words, either α is false or β is true. This means that the sentence $\alpha \to \beta$ is equivalent to the logical sum $(\sim \alpha) \vee \beta$. Hence it follows also by the formulae (17) and (16) that

(18) $$\sim(\alpha \to \beta) = \alpha \wedge (\sim \beta).$$

The operations of addition and multiplication may be extended to a finite number of terms or factors by induction. The generalization of these operations to an arbitrary number (finite or infinite) of terms or factors plays a great role in applications. We perform it in the following way.

Let $\varphi(x)$ be an expression containing a variable x and becoming a sentence (true or false) after substituting in place of x any value of x. Such an expression is called a *propositional function*; e.g. the expression $x > 2$ is a propositional function. It becomes a sentence, when we substitute any real number in place of x; however, it is not a sentence. We say that the values of the variable x for which $\varphi(x)$ is a true sentence, satisfy this propositional function.

We perform the following two operations on the propositional functions:

(19) $$\bigvee_x \varphi(x) \quad \text{and} \quad \bigwedge_x \varphi(x).$$

The first expression is read as follows: "a certain x satisfies the function φ" (that is "there exists an x_0 such that $\varphi(x_0)$ is true"). The second one means that "every x satisfies the function $\varphi(x)$" (that is "for every x the sentence $\varphi(x)$ is true").

As is easily seen, the expressions (19) are sentences; e.g. $\bigvee_x (x > 2)$ is a true sentence and $\bigwedge_x (x > 2)$ is a false sentence. Generally, by adding the operator $\bigwedge_x$ or $\bigvee_x$ (these operators are called the universal and the existential quantifier respectively) before a propositional function we transform the propositional function into a sentence.

Writing the variable x we have to realize the domain of this variable (this domain is often suggested by the symbolism); e.g. for the propositional function $x > 2$, the domain of the variable x is the set of real numbers.

In the case when the domain of the variable is finite, the quantifiers $\bigvee$ and $\bigwedge$ are generalizations of the operations $\vee$ and $\wedge$, respectively. Indeed, if the domain of the variable x consists of n elements $a_1, a_2, ..., a_n$, then

$$\bigvee_x \varphi(x) = [\varphi(a_1) \vee \varphi(a_2) \vee ... \vee \varphi(a_n)],$$

$$\bigwedge_x \varphi(x) = [\varphi(a_1) \wedge \varphi(a_2) \wedge ... \wedge \varphi(a_n)].$$

The de Morgan laws may be generalized to the quantifiers. In this way we have the theorems:

(20) $\quad \sim \bigvee_x \varphi(x) = \bigwedge_x \sim \varphi(x) \quad$ and $\quad \sim \bigwedge_x \varphi(x) = \bigvee_x \sim \varphi(x).$

Indeed, "it is not true that there exists a value of x for which $\varphi(x)$" is a true sentence, if and only if the sentence $\varphi(x)$ is false for every x, that is if the sentence $\sim \varphi(x)$ is true.

The second de Morgan formula may be shown to be true, similarly.

Besides propositional functions of one variable, propositional functions of two or more variables are considered; e.g. $y = x^2$ is a propositional function of two variables (we might denote this function by $\varphi(x, y)$); it becomes a sentence after substituting in place of the variables x and y some values of these variables. Similarly, $z = x + y$ is a propositional function of three variables.

6. SEQUENCES AND SERIES OF FUNCTIONS

Let us now write the definition of the limit of an infinite sequence (cf. § 2.2) by means of these logical symbols:

(21) $\quad (g = \lim\limits_{n=\infty} a_n) \equiv \bigwedge\limits_{\varepsilon} \bigvee\limits_{k} \bigwedge\limits_{n} [(n > k) \to (|a_n - g| < \varepsilon)]$;

here the domain of the variables n and k is the set of positive integers and the domain of the variable ε is the set of positive numbers.

Similarly, the Cauchy condition, necessary and sufficient for the convergence of a sequence $\{a_n\}$ (cf. § 2.7,), is expressed as follows:

(22) $\quad \bigwedge\limits_{\varepsilon} \bigvee\limits_{r} \bigwedge\limits_{n} [(n > r) \to (|a_n - a_r| < \varepsilon)]$.

The definition of the boundedness of a function f in an interval $a \leqslant x \leqslant b$ (cf. § 4.7) is written as follows:

(23) $\quad \bigvee\limits_{M} \bigwedge\limits_{x} |f(x)| < M$.

The (Cauchy) definition of the continuity of a function f at a point x_0 (cf. § 5.2 (3)) may be written in the following way:

(24) $\quad \bigwedge\limits_{\varepsilon} \bigvee\limits_{\delta} \bigwedge\limits_{h} [(|h| < \delta) \to (|f(x_0 + h) - f(x_0)| < \varepsilon)]$.

Here the domain of ε and δ is evidently the set of positive numbers and the domain of the variable h, the set of all real numbers.

Since, according to the definition, a function is continuous in a given set (e.g. for $a < x < b$) when it is continuous at every point of this set, so to write the definition of continuity in a set by means of the logical symbols, one has to precede the expression in the formula (24) by the quantifier $\bigwedge\limits_{x}$ replacing of course x_0 by x. In this way we obtain:

(25) $\quad \bigwedge\limits_{x} \bigwedge\limits_{\varepsilon} \bigvee\limits_{\delta} \bigwedge\limits_{h} [(|h| < \delta) \to (|f(x + h) - f(x)| < \varepsilon)]$.

Before proceeding to further applications of the logical symbols, we note a few simple properties of the quantifiers.

Let a propositional function $\varphi(x, y)$ of two variables be given. It may be easily verified that the formulae

(26) $$\bigwedge_x \bigwedge_y \varphi(x, y) = \bigwedge_y \bigwedge_x \varphi(x, y)$$

and $$\bigvee_x \bigvee_y \varphi(x, y) = \bigvee_y \bigvee_x \varphi(x, y)$$

hold.

Thus, as we see, the order of two universal quantifiers may be changed, as well as the order of two existential quantifiers. However, this is not true in the case of two different quantifiers $\bigvee$ and $\bigwedge$; e.g. changing the order of the quantifiers in the formula (23) expressing the condition of boundedness of the function f we obtain

(27) $$\bigwedge_x \bigvee_M \left(|f(x)| < M\right);$$

but the last condition is satisfied by every function (and so by unbounded functions, too), which is seen substituting $M = |f(x)| + 1$.

So the order of the quantifiers $\bigvee$ and $\bigwedge$ is essential; e.g. it is essential in the condition (23) (for the boundedness of the function f) that the number M does not depend on the variable x; in the condition (27) it is quite opposite: we choose the number M to each x, separately. As is seen, the dependence of a certain quantity on another one is expressed by the order of the quantifiers; thus, it follows directly from the notation. Similarly, in the definition (24) (expressing the continuity of a function f at a point x_0) the order of the quantifiers $\bigwedge_\varepsilon$ and $\bigvee_\delta$ means that δ depends on ε (and not conversely).

We illustrate this situation further by two especially instructive examples, namely, the example of uniform continuity and that of uniform convergence.

First let us note that changing the order of the quantifiers $\bigwedge_x \bigwedge_\varepsilon$ in the formula (25) (expressing the continuity of a function) we obtain an equivalent sentence (since the change concerns the same quantifiers, cf. the formula (26)):

(28) $\bigwedge_{\varepsilon} \bigwedge_{x} \bigvee_{\delta} \bigwedge_{h} \left[(|h| < \delta) \to (|f(x+h) - f(x)| < \varepsilon) \right]$.

However, changing the order of the quantifiers $\bigwedge_{x} \bigvee_{\delta}$ in the formula (28) we obtain a non-equivalent formula, namely

(29) $\bigwedge_{\varepsilon} \bigvee_{\delta} \bigwedge_{x} \bigwedge_{h} \left[(|h| < \delta) \to |f(x+h) - f(x)| < \varepsilon) \right]$,

i.e. the condition of the uniform continuity of the function f in the considered set (cf. § 5.4); but not every function continuous in the interval $a < x < b$ is uniformly continuous (e. g. the function $1/x$ is continuous in the interval $0 < x < 1$ but it is not uniformly continuous in this interval, cf. the remark to Theorem 1 of § 5.4). Although for closed intervals $a \leqslant x \leqslant b$, the usual continuity implies the uniform continuity, but this requires a special proof (to pass from the formulation in Theorem 1, of § 5.4 to the formula (29), x' has to be replaced by $x+h$).

Comparing the formulae (28) and (29) we conclude that they differ only by the order of the quantifiers $\bigwedge_{x}$ and $\bigvee_{\delta}$. However, this difference is very essential: the order $\bigvee_{\delta} \bigwedge_{x}$ (in the formula (29)) expresses the independence of the number δ of the variable x and "uniformity" of the continuity of the function f consists just in this fact.

Now, let us proceed to the notion of the uniform convergence of a sequence of functions (cf. § 6.1). Let a sequence of functions $f_1, f_2, \ldots, f_n, \ldots$ defined in a certain set (e. g. $a < x < b$) be given. Let us assume this sequence to be convergent to a function f. This means by the formula (21) (after substituting $g = f(x)$, $a_n = f_n(x)$) that

(30) $\bigwedge_{x} \bigwedge_{\varepsilon} \bigvee_{k} \bigwedge_{n} \left[(n > k) \to (|f_n(x) - f(x)| < \varepsilon) \right]$.

As is known, the order of the quantifiers $\bigwedge_x \bigwedge_\varepsilon$ may be changed in this formula. However, if we change also the order of the quantifiers $\bigwedge_x$ and $\bigvee_k$, then we obtain the formula for uniform convergence:

(31) $\quad \bigwedge_\varepsilon \bigvee_k \bigwedge_x \bigwedge_n \left[(n > k) \to \left(|f_n(x) - f(x)| < \varepsilon\right)\right].$

For, the order of quantifiers in the formula (31) shows that the number k does not depend on x (but only on ε) and this means that the convergence of the sequence of functions $f_1, f_2, \ldots$ is uniform.

To conclude these remarks we add that sometimes it is convenient to apply the logical symbolism and especially the de Morgan formulae to proofs performed by *reductio ad absurdum*.

For this purpose let us note that the de Morgan formulae (20) may be applied to propositional functions of many variables; they give the following formulae:

(32)
$$\sim \bigvee_x \bigwedge_y \varphi(x, y) = \bigwedge_x \bigvee_y \sim \varphi(x, y),$$
$$\sim \bigwedge_x \bigvee_y \bigwedge_z \varphi(x, y, z) = \bigvee_x \bigwedge_y \bigvee_z \sim \varphi(x, y, z) \quad \text{etc.}$$

For example in the proof of the theorem on the uniform continuity of functions continuous in a closed interval (cf. § 5.4, 1), we use the condition of the uniform continuity given in the following formulation (which does not differ essentially from the formula (29)):

$$\bigwedge_\varepsilon \bigvee_\delta \bigwedge_x \bigwedge_{x'} \left[(|x - x'| < \delta) \to \left(|f(x) - f(x')| < \varepsilon\right)\right].$$

We write the negation of this condition, which gives according to the formulae (32) and (18),

$$\bigvee_\varepsilon \bigwedge_\delta \bigvee_x \bigvee_{x'} \left[(|x - x'| < \delta) \wedge \left(|f(x) - f(x')| \geq \varepsilon\right)\right].$$

The continuation of the proof remains unchanged.

Exercises on § 6

1. Prove that the sum of two uniformly convergent sequences is uniformly convergent. Prove the same for a product, assuming $a \leqslant x \leqslant b$. Show by considering the example $f_n(x) = x\left(1 - \dfrac{1}{n}\right)$, $g_n(x) = \dfrac{1}{x^2}$ that the theorem on the product is false for open intervals.

2. Prove that the radius of convergence r of a power series $\sum\limits_{n=0}^{\infty} a_n x^n$ satisfies the (Cauchy-Hadamard) formula:

$$\frac{1}{r} = \limsup \sqrt[n]{|a_n|}.$$

3. Let a sequence of continuous functions $f_1, f_2, \ldots$ in the interval $a \leqslant x \leqslant b$, be given. Prove that this sequence is uniformly convergent to a function f, if and only if the condition $\lim\limits_{n=\infty} x_n = x$ implies $\lim\limits_{n=\infty} f_n(x_n) = f(x)$.

4. Investigate the uniform convergence of the sequence of functions

$$f_n(x) = x^n(1 - x^n), \quad 0 \leqslant x \leqslant 1$$

and $\quad f_n(x) = \dfrac{1}{nx}, \quad 0 < x \leqslant 1.$

5. Prove that the series

$$S(x) = x(1-x) - x(1-x) + \ldots + x^n(1-x) - x^n(1-x) + \ldots$$

is uniformly and absolutely convergent in the interval $0 \leqslant x \leqslant 1$.

Rearrange the terms of this series in such a way to obtain a nonuniformly convergent series in this interval.

CHAPTER III

DIFFERENTIAL CALCULUS

§ 7. DERIVATIVES OF THE FIRST ORDER

7.1. Definitions [1]

Let a function f be given in an open interval containing the point a. The limit

$$\lim_{h=0} \frac{f(a+h)-f(a)}{h}$$

is called the *derivative of the function f at the point a*. The function

$$g(h) = \frac{f(a+h)-f(a)}{h},$$

the limit of which is considered by h tending to 0, is called the *difference quotient of the function f at the point a for the increment h*.

The derivative is written symbolically as follows:

(1) $$\left[\frac{df(x)}{dx}\right]_{x=a} = \lim_{h=0} \frac{f(a+h)-f(a)}{h}$$

or more briefly, $f'(a)$. Yet more briefly, writing $y = f(x)$ we denote the derivative of a function f at an arbitrary point x by $\frac{dy}{dx}$ (the derivative "dy over dx"). By this symbolism (originating with Leibniz), the derivative is represented as a quotient of two differentials dy and dx (cf. § 7.13). However, the symbol $\frac{dy}{dx}$ must be treated

---

[1] The differential and integral calculus was created by Newton and Leibniz on the end of the 17th century.

as a whole, attributing to the differentials no separate mathematical significance.

Geometrically, the derivative is interpreted as follows. Let a curve $y = f(x)$ be given. Let us draw a straight line through the points $[a, f(a)]$ and $[a+h, f(a+h)]$, h being a fixed positive value. This straight line is called a *secant* with regard to the given curve. It is easily seen that the difference quotient $g(h)$ is the tangent of the angle a_h between the secant directed by the increasing abscissae and the positive direction of the X-axis. The

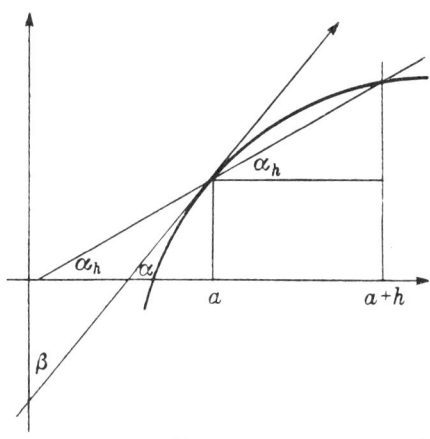

Fig. 11

limiting position to which the secant tends as h tends to 0 will be considered as the position of the tangent. Thus,

(2) $$f'(a) = \tan a,$$

where a is the angle between the positive direction of the tangent to the curve $y = f(x)$ at the point a and the positive direction of the X-axis.

Not every continuous function possesses a derivative, i. e. not every curve possesses a tangent at any point. E. g. the function $f(x) = |x|$ does not possess a derivative at the point 0, for in this case

$$\lim_{h=+0} \frac{f(h)-f(0)}{h} = 1 \quad \text{and} \quad \lim_{h=-0} \frac{f(h)-f(0)}{h} = -1.$$

In this case we may speak about *one-side* derivatives which are equal to 1 and −1, respectively. Generally, a right-side or left-side derivative is the right-side or left-side limit:

(3) $f'_+(a) = \lim\limits_{h=+0} \dfrac{f(a+h)-f(a)}{h}$, $f'_-(a) = \lim\limits_{h=-0} \dfrac{f(a+h)-f(a)}{h}$,

respectively.

However, there exist continuous functions which do not even have a one-side derivative. It is easily seen that the function f defined by the conditions

(4) $\qquad f(x) = x \sin \dfrac{1}{x} \quad$ for $\quad x \neq 0, \quad f(0) = 0$

is continuous but has no derivative at the point 0.

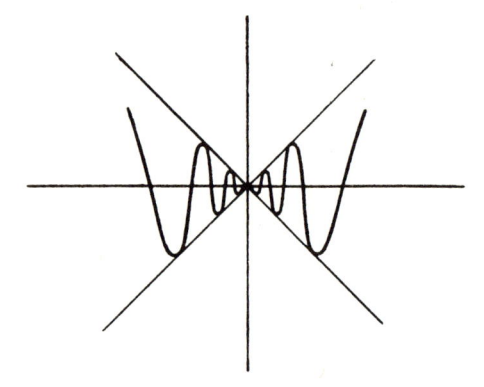

Fig. 12

For, substituting for h the two sequences of values

$$\dfrac{2}{\pi}, \dfrac{2}{5\pi}, \dfrac{2}{9\pi}, \ldots \quad \text{and} \quad \dfrac{2}{3\pi}, \dfrac{2}{7\pi}, \dfrac{2}{11\pi}, \ldots,$$

we get in the limit 1 in the first case and in the second one, −1.

The non-existence of the derivative $f'(0)$ follows also from the non-existence of the limit $\lim\limits_{x=0} \sin \dfrac{1}{x}$, since we should have

$$f'(0) = \lim_{h=0} \frac{f(h)-f(0)}{h} = \lim_{h=0} \sin \frac{1}{h}$$

(cf. § 4 (8)).

There exist still more singular continuous functions, namely functions which do not have a derivative at any point; thus, they represent continuous curves which do not have a tangent at any point.

Besides finite derivatives we consider also infinite derivatives. For example we shall prove that

$$\left(\frac{d\sqrt[3]{x}}{dx}\right)_{x=0} = \infty, \quad \left(\frac{d\sqrt{x}}{dx}\right)_{x=+0} = \infty.$$

Indeed,

$$\left(\frac{d\sqrt[3]{x}}{dx}\right)_{x=0} = \lim_{h=0} \frac{\sqrt[3]{h}}{h} = \lim_{h=0} \frac{1}{h^{2/3}} = \infty,$$

for $\lim_{h=0} h^{2/3} = 0$. Analogously $\lim_{h=+0} \sqrt{h} = +0$, whence the second formula follows.

A function is called *differentiable in an open interval* if it has a finite derivative at any point of this interval; in saying that a function is differentiable in a closed interval $a \leqslant x \leqslant b$ we assume that it has a derivative at any point inside this interval and one-side derivatives at the ends of this interval.

Similarly, the assumption of continuity of the derivative $f'(x)$ in the interval $a \leqslant x \leqslant b$ means that this derivative is continuous inside the interval ab, the right-side derivative is right-side continuous at the point a and the left-side derivative is left-side continuous at b.

By a *normal* to a curve $y = f(x)$ at a point $[x, f(x)]$ we understand a straight line perpendicular to the tangent at this point and passing through this point.

According to the geometrical interpretation of the derivative given above, the tangent to a curve at a point

(x, y), where $y = f(x)$, is expressed by the equation

(5) $\quad \dfrac{Y-y}{X-x} = f'(x) \quad \text{or} \quad X = x \quad \text{if} \quad f'(x) = \infty;$

X and Y denote the coordinates on the tangent.

Thus, the equation of the normal to a curve at a point (x, y) is:

(6) $\quad \dfrac{X-x}{Y-y} = -f'(x) \quad \text{or} \quad Y = y, \quad \text{if} \quad f'(x) = \infty.$

Besides the geometrical interpretation the derivative has also important interpretations in physics. In particular, the velocity of the point moving along a straight line is expressed as the derivative of the distance with respect to the time: $v = \dfrac{ds}{dt}$, where the distance s travelled is expressed as a function of the time: $s = f(t)$. Thus, the velocity at a moment t is the limit of the mean velocity in the time from t to $t+h$, if h tends to 0; for the differences quotient is just this mean velocity.

7.2. Differentiation of elementary functions

First let us note that

1. *If a function f is differentiable at a point x, then it is continuous at this point.*

Indeed, since the limit $\lim\limits_{h=0} \dfrac{f(x+h)-f(x)}{h}$ exists by assumption (and is finite), we have

$$\lim_{h=0}[f(x+h)-f(x)] = \lim_{h=0}\dfrac{f(x+h)-f(x)}{h} \cdot \lim_{h=0} h = 0,$$

whence

$$\lim_{h=0} f(x+h) = f(x),$$

which means that the function f is continuous at the point x.

2. *Let a function f have a constant value: $f(x) = c$.* Then $\dfrac{dc}{dx} = 0.$

7. DERIVATIVES OF THE FIRST ORDER

For, we have $f(x+h) = c = f(x)$, whence
$$\lim_{h=0} \frac{f(x+h)-f(x)}{h} = 0.$$

3. $\dfrac{dx}{dx} = 1,$

for $\lim\limits_{h=0} \dfrac{(x+h)-x}{h} = \lim\limits_{h=0} \dfrac{h}{h} = 1.$

The following formulae concern the differentiation of a sum, a difference, a product and a quotient of two functions $y = f(x)$ and $z = g(x)$, differentiable at a point x:

4. $\dfrac{d(y \pm z)}{dx} = \dfrac{dy}{dx} \pm \dfrac{dz}{dx}.$ 5. $\dfrac{d(yz)}{dx} = y \dfrac{dz}{dx} + z \dfrac{dy}{dx}.$

6. $\dfrac{d(y/z)}{dx} = \dfrac{z \dfrac{dy}{dx} - y \dfrac{dz}{dx}}{z^2},$

whence

6'. $\dfrac{d(1/z)}{dx} = -\dfrac{1}{z^2} \cdot \dfrac{dz}{dx}$ (if $z \neq 0$).

The proof of Formula 4 follows from the equations
$$\frac{d(y+z)}{dx} = \lim_{h=0} \frac{[f(x+h)+g(x+h)]-[f(x)+g(x)]}{h}$$
$$= \lim_{h=0} \frac{f(x+h)-f(x)}{h} + \lim_{h=0} \frac{g(x+h)-g(x)}{h} = \frac{dy}{dx} + \frac{dz}{dx}.$$

The proof for the difference is analogous.

In the proof of Formula 5 we shall apply the continuity of the function f at the point x (cf. 1); thus we have $\lim\limits_{h=0} f(x+h) = f(x)$. Now,

$$\frac{d(yz)}{dx} = \lim_{h=0} \frac{f(x+h) \cdot g(x+h) - f(x) \cdot g(x)}{h}$$
$$= \lim_{h=0} \frac{[f(x+h) \cdot g(x+h) - f(x+h) \cdot g(x)]}{h}$$
$$+ \frac{[f(x+h) \cdot g(x) - f(x) \cdot g(x)]}{h}$$

$$= \lim_{h=0} f(x+h) \cdot \lim_{h=0} \frac{g(x+h)-g(x)}{h} + g(x) \cdot \lim_{h=0} \frac{f(x+h)-f(x)}{h}$$

$$= y \frac{dz}{dx} + z \frac{dy}{dx}.$$

To prove 6, we shall prove first 6'.

$$\frac{d(1/z)}{dx} = \lim_{h=0} \frac{1}{h} \left[\frac{1}{g(x+h)} - \frac{1}{g(x)} \right]$$

$$= -\frac{1}{\lim_{h=0} g(x+h)} \cdot \frac{1}{g(x)} \cdot \lim_{h=0} \frac{g(x+h)-g(x)}{h} = -\frac{1}{z^2} \cdot \frac{dz}{dx}$$

for $\lim_{h=0} g(x+h) = g(x) = z$ (moreover, since $g(x) \neq 0$, there exists a $\delta > 0$ such that $g(x+h) \neq 0$ for $|h| < \delta$).

Formulae 5 and 6' imply 6:

$$\frac{d(y/z)}{dx} = \frac{d}{dx}\left(y \cdot \frac{1}{z}\right) = y \cdot \frac{d}{dx}\left(\frac{1}{z}\right) + \frac{1}{z} \frac{dy}{dx} = \frac{1}{z^2}\left(z \frac{dy}{dx} - y \frac{dz}{dx}\right)$$

(in the first of the above equations we apply the notation $\frac{d}{dx} F(x)$ in place of $\frac{dF(x)}{dx}$).

Substituting $g(x) = c$ in formulae 4 and 5 we obtain immediately:

4'. $\dfrac{d(y+c)}{dx} = \dfrac{dy}{dx}$,

5'. $\dfrac{d(cy)}{dx} = c \dfrac{dy}{dx}$.

The first of these formulae means that a translation of a curve, parallel to the Y-axis has no influence on the value of the angle between the tangent and the X-axis; the second one means that a change of the scale on the Y-axis influences the tangens of the considered angle in the same proportion.

Applying Formula 5 and the principle of induction we prove easily that for integer exponents $n \geqslant 2$ the formula

7. $\dfrac{d(x^n)}{dx} = nx^{n-1}$

holds.

It is immediately seen that this formula holds also for $n = 0$ and for $n = 1$, assuming that if $x = 0$, then in the case $n = 0$ the right-side of the formula has to be replaced by 0 and in the case $n = 1$, by 1 (as Formulae 2 and 3 show directly).

This formula remains also true for negative integers n (when $x \neq 0$). Namely, we have in this case, by 6' and 7:

$$\frac{d(x^n)}{dx} = \frac{1}{dx}\left(\frac{1}{x^{-n}}\right) = -\frac{1}{x^{-2n}} \cdot \frac{d(x^{-n})}{dx} = \frac{n}{x^{-2n}} x^{-n-1} = nx^{n-1}.$$

Later we shall generalize formula 7 to arbitrary real exponents (for $x > 0$).

From the formulae already proved we deduce easily the formula for the derivative of a polynomial:

8. $\dfrac{d}{dx}(a_0 + a_1 x + \ldots + a_n x^n) = a_1 + 2a_2 x + 3a_3 x^2 + \ldots + n a_n x^{n-1}.$

We shall now calculate the derivatives of the trigonometric functions.

9. $\dfrac{d \sin x}{dx} = \cos x.$

10. $\dfrac{d \cos x}{dx} = -\sin x.$

11. $\dfrac{d \tan x}{dx} = \dfrac{1}{\cos^2 x}.$

We shall base the proof of formula 9 on the formula for the difference of sine known from trigonometry and on the formula $\lim\limits_{h=0} \dfrac{\sin h}{h} = 1$ proved in § 4.5, (5). Now,

$$\frac{d \sin x}{dx} = \lim_{h=0} \frac{1}{h}[\sin(x+h) - \sin x] = \lim_{h=0} \frac{2}{h} \cdot \sin\frac{h}{2} \cdot \cos\left(x + \frac{h}{2}\right)$$

$$= \lim_{h=0} \frac{\sin(h/2)}{(h/2)} \cdot \lim_{h=0} \cos\left(x + \frac{h}{2}\right) = \cos x$$

by virtue of the continuity of the function $\cos x$.

Similarly,

$$\frac{d\cos x}{dx} = \lim_{h=0}\frac{1}{h}[\cos(x+h)-\cos x] = \lim_{h=0}\frac{-2}{h}\cdot\sin\frac{h}{2}\cdot\sin\left(x+\frac{h}{2}\right)$$

$$= -\lim_{h=0}\frac{\sin(h/2)}{(h/2)}\cdot\lim_{h=0}\sin\left(x+\frac{h}{2}\right) = -\sin x.$$

According to 6 Formulae 9 and 10 imply Formula 11, for we have

$$\frac{d\tan x}{dx} = \frac{d}{dx}\left(\frac{\sin x}{\cos x}\right) = \frac{1}{\cos^2 x}\left(\cos x\cdot\frac{d\sin x}{dx} - \sin x\cdot\frac{d\cos x}{dx}\right)$$

$$= \frac{1}{\cos^2 x}(\cos^2 x + \sin^2 x) = \frac{1}{\cos^2 x}.$$

12. $\dfrac{d\log x}{dx} = \dfrac{1}{x}$. More generally, $\dfrac{d\log_a x}{dx} = \dfrac{1}{x\log a}$.

Indeed, according to the general properties of logarithms (§ 5.4, (10) and (11)) we have:

$$\frac{1}{h}[\log(x+h) - \log x] = \frac{1}{h}\log\left(1+\frac{h}{x}\right) = \frac{1}{x}\log\left[\left(1+\frac{h}{x}\right)^{\frac{x}{h}}\right].$$

Let us substitute $y = \dfrac{h}{x}$. Since $\lim_{h=0} y = 0$, we have (cf. § 4, (16)):

$$\lim_{h=0}\frac{1}{h}[\log(x+h)-\log x] = \frac{1}{x}\lim_{h=0}\log\left[\left(1+\frac{h}{x}\right)^{\frac{x}{h}}\right]$$

$$= \frac{1}{x}\lim_{y=0}\log(1+y)^{\frac{1}{y}}.$$

Moreover, since $\lim_{y=0}(1+y)^{\frac{1}{y}} = e$ (cf. § 4, (19)) and since $\log z$ is a continuous function at the point $z = e$, we have (cf. § 5, (6)):

$$\lim_{y=0}\log(1+y)^{\frac{1}{y}} = \log\lim_{y=0}(1+y)^{\frac{1}{y}} = \log e = 1.$$

Thus

$$\frac{d\log x}{dx} = \lim_{h=0} \frac{1}{h}[\log(x+h) - \log x] = \frac{1}{x}.$$

The second part of Formula 12 follows according to Formula 5' from the first one:

$$\frac{d}{dx}\log_a x = \frac{d}{dx}\left(\frac{\log x}{\log a}\right) = \frac{1}{\log a} \cdot \frac{d\log x}{dx} = \frac{1}{x\log a}.$$

Further formulae on differentiation of elementary functions will be deduced from the general formula on the derivative of an inverse function.

7.3. Differentiation of inverse functions

Let a differentiable and one-to-one function $y = f(x)$ be given in an interval $a \leqslant x \leqslant b$. As it is well known (§ 5.5, 1 and 2) there exists a function $x = g(y)$ inverse to the given one and continuous in the interval $f(a) \leqslant y \leqslant f(b)$ or $f(b) \leqslant y \leqslant f(a)$, respectively (depending on that whether the function f is increasing or decreasing). We shall prove that the function g is differentiable in this interval. Namely,

1. $\dfrac{dx}{dy} = 1 : \dfrac{dy}{dx}, \quad if \quad \dfrac{dy}{dx} \neq 0.$

Given x, let us write $k = f(x+h) - f(x)$. Then $f(x+h) = y+k$, i.e. $x+h = g(y+k)$, whence $h = g(y+k) - g(y)$. With variable k, the increment h is a function of k. By the continuity of the function g, we have $\lim_{k=0} h = 0$; moreover, we have $h \neq 0$ for $k \neq 0$, since g is one-to-one. Applying formula (16) of § 4 we obtain:

$$\frac{dx}{dy} = \lim_{k=0} \frac{g(y+k) - g(y)}{k} = \lim_{k=0} \frac{h}{f(x+h) - f(x)}$$

$$= \lim_{h=0} \frac{h}{f(x+h) - f(x)} = \frac{1}{\dfrac{dy}{dx}}.$$

Remarks. (α) As at the ends of the interval of y, i.e. at the points $f(a)$ and $f(b)$, one-side derivatives exist, then our formula becomes of the following form:

1'. $\quad \dfrac{dx}{dy_\pm} = 1 : \dfrac{dy}{dx_\pm},$

where the signs are the same for increasing functions and opposite for decreasing functions.

For if the function f is increasing, then the function g is also increasing and the increments h and k have the same sign and consequently, $\lim\limits_{k=+0} h = +0$ and $\lim\limits_{k=-0} h = -0$. Yet if the function f is decreasing, then $\lim\limits_{k=+0} h = -0$ and $\lim\limits_{k=-0} h = +0$.

(β) Geometrically, Theorem 1 may be illustrated as follows.

Let us denote by α the angle between the tangent to the curve and the X-axis and by β the angle between this tangent and the Y-axis (cf. Fig. 11). Then $\tan \beta = \cot \alpha$, i.e. $\tan \beta = 1/\tan \alpha$ according to Formula 1.

(γ) Formula 1 exhibits the convenience of the Leibniz symbolism: in passing to the inverse function the derivative $\dfrac{dy}{dx}$ behaves like a fraction.

Now, we shall give some applications of Formula 1 in the differentiation of elementary functions.

2. $\dfrac{de^x}{dx} = e^x$. More generally: 2'. $\dfrac{da^x}{dx} = a^x \log a$.

Let us write $y = e^x$, i.e. $x = \log y$. Since, by 12, $\dfrac{dx}{dy} = \dfrac{1}{y}$, we have $\dfrac{dy}{dx} = 1 : \dfrac{dx}{dy} = y$. Substituting $y = e^x$ we obtain Formula 2.

The proof of Formula 2' is completely analogous.

3. $\dfrac{d \arcsin x}{dx} = \dfrac{1}{\sqrt{1-x^2}}$.

4. $\dfrac{d \arccos x}{dx} = \dfrac{-1}{\sqrt{1-x^2}}$.

5. $\dfrac{d \arctan x}{dx} = \dfrac{1}{1+x^2}$.

To prove 3, let us write $y = \arcsin x$, i.e. $x = \sin y$. Then we have

$$\frac{dx}{dy} = \cos y = \sqrt{1-\sin^2 y} = \sqrt{1-x^2}$$

(where the root has the sign $+$ since $-\tfrac{1}{2}\pi \leqslant y \leqslant \tfrac{1}{2}\pi$ according to the definition of the function $\arcsin x$, cf. § 4.4).

Hence $\dfrac{dy}{dx} = \dfrac{1}{\sqrt{1-x^2}}$.

Similarly, writing $y = \arccos x$, i.e. $x = \cos y$, we have

$$\frac{dx}{dy} = -\sin y = -\sqrt{1-\cos^2 y} = -\sqrt{1-x^2},$$

whence $\dfrac{dy}{dx} = \dfrac{-1}{\sqrt{1-x^2}}$.

Finally, if $y = \arctan x$, then $x = \tan y$ and so

$$\frac{dx}{dy} = \frac{1}{\cos^2 y} = 1 + \tan^2 y = 1 + x^2, \quad \text{whence} \quad \frac{dy}{dx} = \frac{1}{1+x^2}.$$

Before we proceed to formulae concerning the superpositions of the elementary functions considered above we shall prove some general theorems on the derivatives.

7.4. Extrema of functions. Rolle theorem

Let a function f be defined in a neighbourhood of a point a (i.e. in a certain interval containing this point inside). If there exists a $\delta > 0$ such that the inequality $|h| < \delta$ implies the inequality

(7) $$f(a+h) \leqslant f(a),$$

then we say that *the function f has a maximum at the point a.*

Further if the inequality $|h| < \delta$ implies

(8) $$f(a+h) \geqslant f(a),$$

then we say that *the function f has a minimum at the point a.*

In other words, a point a is a maximum (or a minimum) of the function f, if such an interval surrounding the point a exists that $f(a)$ is the greatest (or the least) among all values of the function f in this interval.

Replacing the signs $\leqslant$ and $\geqslant$ in formulae (7) and (8) by the signs $<$ and $>$, we obtain the *proper* maximum and minimum, respectively.

The maxima and minima have the common name *extrema*.

EXAMPLES. The function x^2 has a minimum at the point 0; the function $\sin x$ assumes its maximum and minimum values at points being odd multiples of $\tfrac{1}{2}\pi$, alternately.

As regards the relation between the notion of the extremum of a function and the upper and lower bounds of this function we note first of all that the notion of the extremum is a local notion and the notion of the bound of a function is an integral notion: we speak about the extremum of a function at a given point and about the bound of a function in a given interval. To state whether a function has an extremum at a point a it suffices to know the values of this function in an arbitrary neighbourhood of the point a. On the other hand, to find the upper bound of a function in an interval we must know the behaviour of this function in the whole interval.

It follows immediately from the definition that

1. *If a function $f(x)$, $a \leqslant x \leqslant b$, attains its upper bound at a point c lying inside the interval ab (i.e. $a < c < b$), then the function has also a maximum at this point.*

An analogous theorem concerns the lower bound and the minimum.

Yet if the upper bound of the function is attained at one of the ends of the interval ab, then it is not a maximum of this function, since the function is not defined in any neighbourhood of the ends of this interval. E.g. the function $y = x$ considered in the interval $0 \leqslant x \leqslant 1$ attains its upper bound at the point 1; however, this is not a maximum.

2. *If a function f is differentiable at a point c and has an extremum at this point, then $f'(c) = 0$.*

Let us assume the function to possess a maximum at the point c (in the case of the minimum, the arguments are analogous). Let a number $\delta > 0$ be chosen in such a way that the inequality $f(c+h) - f(c) \leqslant 0$ holds for $|h| < \delta$. Thus

$$\frac{f(c+h) - f(c)}{h} \leqslant 0 \text{ for } h > 0$$

and $\quad \dfrac{f(c+h) - f(c)}{h} \geqslant 0 \text{ for } h < 0.$

Since the derivative $f'(c)$ exists by assumption, we have

$$f'_+(c) = f'(c) = f'_-(c).$$

At the same time it follows from the previous inequalities: $f'_+(c) \leqslant 0 \leqslant f'_-(c)$. Hence

$$f'_+(c) = 0 = f'_-(c), \quad \text{i.e.} \quad f'(c) = 0.$$

Remark. The converse theorem does not hold: the equality $f'(c) = 0$ may be satisfied although the function does not possess an extremum at the point c. It is so e.g. in the case of the function x^3 at the point 0.

Geometrically, the existence of an extremum of the function f at a point c means (in the case when the function is differentiable), that the tangent to the curve $y = f(x)$ at the point $[c, f(c)]$ is parallel to the X-axis (or is identical with the X-axis).

3. ROLLE THEOREM. *Let the function f be continuous in a closed interval $a \leqslant x \leqslant b$ and differentiable inside this*

interval. If $f(a) = f(b)$, then a c exists such that $a < c < b$ and $f'(c) = 0$.

Indeed, if the function f is constant, then we have $f'(x) = 0$ for every x lying between a and b; hence, in this case we may take any value of x as c.

Thus, let us assume that the function f is not constant everywhere, e.g. let us assume that f has values $> f(a)$. Denoting by M the upper bound of this function, we have then $M > f(a)$. According to the Weierstrass theorem (§ 5.4, 2) there exists a c in the interval ab such that $f(c) = M$. Moreover, $a \neq c \neq b$, since $f(a) = f(b)$ by assumption. Thus $a < c < b$. This means that the function f attains its upper bound at a point c lying inside the interval ab; according to Theorem 1, the function f possesses at this point a maximum, and by Theorem 2, we have $f'(c) = 0$.

Rolle theorem is easily understood geometrically: if a curve (having a tangent at every point) crosses the X-axis in two points, then at a certain point the tangent to this curve is parallel to the X-axis.

Remark. Rolle theorem may be expressed also in the following form: if $f(x) = f(x+h)$, then a Θ exists such that

(9) $\qquad f'(x + \Theta h) = 0, \quad 0 < \Theta < 1;$

here the same assumptions on the continuity and the differentiability of $f(x)$ are made as in Rolle theorem (yet we do not assume that $h > 0$ but only that $h \neq 0$).

7.5. Lagrange [1] and Cauchy theorems

Let us assume as in the Rolle theorem that the function f is continuous in the interval $a \leqslant x \leqslant b$ and differentiable inside this interval. Then we have the

---

[1] Joseph Louis Lagrange (1736-1813), the most eminent mathematician of the 18th century, one of the authors of the theory of differential equations.

7. DERIVATIVES OF THE FIRST ORDER

(Lagrange) *mean-value theorem* (called also the theorem of "finite increments"):

1. $\dfrac{f(b)-f(a)}{b-a} = f'(a+\Theta h)$,

where $h = b-a$ and Θ is a suitably chosen number such that $0 < \Theta < 1$.

Before proceeding to the proof we interpret the above formula geometrically. Its left side means the tangent of the angle α between the X-axis and the straight line joining the points $[a, f(a)]$ and $[b, f(b)]$. Hence the theorem

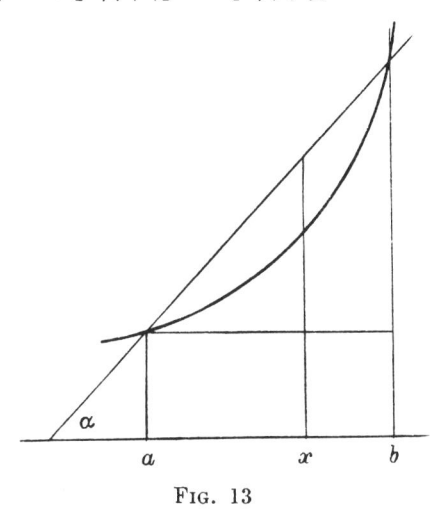

Fig. 13

states that there exists a tangent to the curve which is parallel to this straight line (the "secant to the curve"). A special case gives Rolle theorem, namely, when this secant is parallel to the X-axis. We shall reduce the proof of the Lagrange theorem to this theorem.

Let us draw a straight line parallel to the Y-axis through an arbitrary point x lying between a and b (cf. Fig. 13) and let us denote by $g(x)$ the length of the segment of this straight line contained between the curve and the above defined secant. Thus the length of the segment contained between the X-axis and the secant is

$g(x)+f(x)$. On the other hand, this length is $f(a)+(x-a)\tan\alpha$, whence

(10) $$g(x) = f(a) - f(x) + (x-a)\frac{f(b)-f(a)}{b-a}.$$

The function g satisfies the assumptions of Rolle theorem. It is continuous in the interval $a \leqslant x \leqslant b$, differentiable:

$$g'(x) = -f'(x) + \frac{f(b)-f(a)}{b-a},$$

and $g(a) = 0 = g(b)$. Hence the function $g'(x)$ vanishes at a certain point between a and b. In other words, a Θ exists such that $0 < \Theta < 1$ and that

$$g'(a+\Theta h) = 0, \quad \text{i.e.} \quad 0 = -f'(a+\Theta h) + \frac{f(b)-f(a)}{b-a}.$$

So we have obtained Formula 1.

Remarks. (α) The geometrical considerations have been only of a heuristic character here; they make clear why the function g is introduced. However, the proof of the Lagrange theorem might begin with the definition of the function g by means of the formula (10).

(β) Analogously to the formula (9), the equation 1 may be also written in the following form:

(11) $$f(x+h) = f(x) + h \cdot f'(x+\Theta h).$$

Lagrange theorem implies the following two corollaries, being of fundamental importance in the integral calculus.

3. *If $f'(x) = 0$ for every x lying inside the interval ab, then the function f has a constant value in this interval.*

For, according to (11) we have $f(x+h) = f(x)$ for any x and h and this means that f is of a constant value.

4. *If we have $f'(x) = g'(x)$ for all $a < x < b$, then $f(x) = g(x)+$ constans, i.e. the functions f and g differ by a constant.*

Indeed, $[f(x)-g(x)]' = f'(x)-g'(x) = 0$ and this means that the function $f(x)-g(x)$ has the derivative $= 0$ everywhere. Hence, by Theorem 3, this function is constant. Writing $f(x)-g(x) = C$ we have $f(x) = g(x)+C$.

Remark. Making in Theorem 3 the additional assumption that the function f is continuous in the whole interval $a \leqslant x \leqslant b$, this function is of constant value also in this whole interval. Namely, we may substitute $x = a$ in the proof.

A similar remark concerns Theorem 4.

Lagrange theorem may be generalized as follows:

5. CAUCHY THEOREM. *If the functions f and f_1 are continuous in the whole interval $a \leqslant x \leqslant b$ and differentiable inside this interval and if $f_1'(x) \neq 0$ for any x, then*

(12) $$\frac{f(b)-f(a)}{f_1(b)-f_1(a)} = \frac{f'(a+\Theta h)}{f_1'(a+\Theta h)}, \quad 0 < \Theta < 1,$$

where $h = b-a$.

Lagrange theorem is obtained from Cauchy theorem by making the substitution $f_1(x) = x$. Conversely, to prove Cauchy theorem let us replace the function $g(x)$ in the formula (10) by the following function:

$$g_1(x) = f(a)-f(x)+[f_1(x)-f_1(a)]\frac{f(b)-f(a)}{f_1(b)-f_1(a)}.$$

(Let us note that $f_1(b) \neq f_1(a)$ according to the assumption $f_1'(x) \neq 0$ and Rolle theorem.)

This function satisfies the assumption of Rolle theorem:

(13) $$g_1'(x) = -f'(x)+f_1'(x)\frac{f(b)-f(a)}{f_1(b)-f_1(a)}, \quad g_1(a) = 0 = g_1(b).$$

Thus there exists a Θ between 0 and 1 such that $g'(a+\Theta h) = 0$. Substituting $x = a+\Theta h$ in the formula (13), we obtain equation (12).

Analogously to the formulae (9) and (11) we obtain from Cauchy theorem:

(14) $$\frac{f(x+h)-f(x)}{f_1(x+h)-f_1(x)} = \frac{f'(x+\Theta h)}{f_1'(x+\Theta h)}.$$

Remark. It is essential in the Cauchy formula that Θ indicates the same number in the numerator and in the denominator. A direct application of Formula 1 to the function f and to the function f_1 and the consideration of the quotient of the obtained expressions leads (in most cases) to different numbers Θ in the numerator and in the denominator.

7.6. Differentiation of composite functions

Let $y = f(x)$, $z = g(y)$, where the function g is defined for the values y of the function f, the functions f and g are differentiable and the derivative g' is continuous. The following formula gives the derivative of the composite function $g[f(x)]$ in terms of the derivatives f' and g'.

1. $\dfrac{dz}{dx} = \dfrac{dz}{dy} \cdot \dfrac{dy}{dx}$, i.e. [1] $\dfrac{dgf(x)}{dx} = \left[\dfrac{dg(y)}{dy}\right]_{y=f(x)} \cdot \dfrac{df(x)}{dx}.$

Given x and $h \neq 0$, let us write $k = f(x+h)-f(x)$, i.e. $f(x+h) = y+k$.

Applying the formula (11) on the mean-value to the function g we obtain:

$$\frac{gf(x+h)-gf(x)}{h} = \frac{g(y+k)-g(y)}{h} = g'(y+\Theta k) \cdot \frac{k}{h}$$
$$= g'(y+\Theta k) \cdot \frac{f(x+h)-f(x)}{h}.$$

To get Formula 1 we have to pass to the limit as h tends to 0. Now, by virtue of the continuity of the function g we have $\lim\limits_{h=0} k = 0$. Since $0 < \Theta < 1$, this also implies $\lim\limits_{h=0} \Theta k = 0$ (let us note that Θ is also a function of the variable h). Thus $\lim\limits_{h=0}(y+\Theta k) = y$, which

---
[1] For brevity, we write $gf(x)$ instead of $g[f(x)]$.

gives $\lim\limits_{h=0} g'(y+\Theta k) = g'(y) = g'[f(x)]$ according to the continuity of the function g'. Consequently,

$$\frac{dgf(x)}{dx} = \lim_{h=0}\frac{gf(x+h)-gf(x)}{h}$$
$$= \lim_{h=0}g'(y+\Theta k)\lim_{h=0}\frac{f(x+h)-f(x)}{h} = g'[f(x)]\cdot f'(x).$$

Remark. The theorem is also true without the assumption of the continuity of the function g'. However, then it requires another more complicated proof.

Applications. Formula 1 enables us to generalize Formula 7 of § 7.2 in the following way: for any real a and any positive x, we have

2. $\dfrac{dx^a}{dx} = a \cdot x^{a-1}$.

Namely, we have $x^a = e^{a\log x}$ and so substituting $z = e^y$ and $y = a \cdot \log x$ we find

$$\frac{dx^a}{dx} = \frac{de^y}{dy}\cdot\frac{d(a\log x)}{dx} = e^y\cdot\frac{a}{x} = x^a\cdot\frac{a}{x} = a\cdot x^{a-1}.$$

The above calculation could be written also in the following way:

$$\frac{dx^a}{dx} = \frac{de^{a\log x}}{d(a\cdot\log x)}\cdot\frac{d(a\cdot\log x)}{dx} = e^{a\log x}\cdot\frac{a}{x} = x^a\cdot\frac{a}{x} = a\cdot x^{a-1}.$$

3. $\dfrac{dx^x}{dx} = x^x(\log x+1)$,

for

$$\frac{dx^x}{dx} = \frac{de^{x\log x}}{dx} = \frac{de^{x\log x}}{d(x\cdot\log x)}\cdot\frac{d(x\cdot\log x)}{dx}$$
$$= e^{x\log x}(\log x+1) = x^x(\log x+1).$$

Let us note that formula 2 of § 7.3, $\dfrac{da^x}{dx} = a^x\log a$, proved previously, could be proved in an analogous way, because $a^x = e^{x\log a}$.

In practice we sometimes apply Formula 1 several times; e.g. if $y = f(x)$, $z = g(y)$, $w = h(z)$, we calculate the derivative $\dfrac{dh\{g[f(x)]\}}{dx}$, applying the equality:

$$\frac{dw}{dx} = \frac{dw}{dz} \cdot \frac{dz}{dy} \cdot \frac{dy}{dx}.$$

As seen, the derivatives in the above formula behave like ordinary fractions.

4. $\dfrac{d \log f(x)}{dx} = \dfrac{f'(x)}{f(x)}.$

This expression is called the *logarithmic derivative* of the function f. A knowledge of the logarithmic derivative gives the usual derivative at once. This is easy to show by considering the example of the function $y = x^x$ (where the $\log y = x \log x$ is differentiated).

EXAMPLES.

(α) $\dfrac{d \log \sin (x^2)}{dx} = \dfrac{d \log \sin (x^2)}{d \sin (x^2)} \cdot \dfrac{d \sin (x^2)}{dx^2} \cdot \dfrac{dx^2}{dx}$

$= \dfrac{1}{\sin (x^2)} \cos (x^2) 2x = 2x \cot (x^2).$

(β) *The derivatives of the hyperbolic functions.* By the hyperbolic sine and cosine we understand the functions

(15) $\quad \sinh x = \dfrac{e^x - e^{-x}}{2} \quad$ and $\quad \cosh x = \dfrac{e^x + e^{-x}}{2}.$

To find their derivatives let us note that

$$\frac{de^{-x}}{dx} = \frac{de^{-x}}{d(-x)} \frac{d(-x)}{dx} = -e^{-x}.$$

Hence we find easily:

(16) $\quad \dfrac{d \sinh x}{dx} = \cosh x, \quad \dfrac{d \cosh x}{dx} = \sinh x.$

Now let us write

(17) $\quad \tanh x = \dfrac{\sinh x}{\cosh x} = \dfrac{e^x - e^{-x}}{e^x + e^{-x}}.$

7. DERIVATIVES OF THE FIRST ORDER

Since, as it is easily seen,

(18) $$\cosh^2 x - \sinh^2 x = 1,$$

we have

(19) $$\frac{d\tanh x}{dx} = \frac{d\frac{\sinh x}{\cosh x}}{dx} = \frac{\cosh^2 x - \sinh^2 x}{\cosh^2 x} = \frac{1}{\cosh^2 x}.$$

The functions $\sinh x$ and $\cosh x$ have inverse functions. Namely, let $y = \frac{e^x - e^{-x}}{2}$. Then $e^x - e^{-x} - 2y = 0$, i.e. $e^{2x} - 2ye^x - 1 = 0$, whence

$$e^x = y + \sqrt{y^2 + 1}, \quad \text{i.e.} \quad x = \log\left(y + \sqrt{y^2 + 1}\right)$$

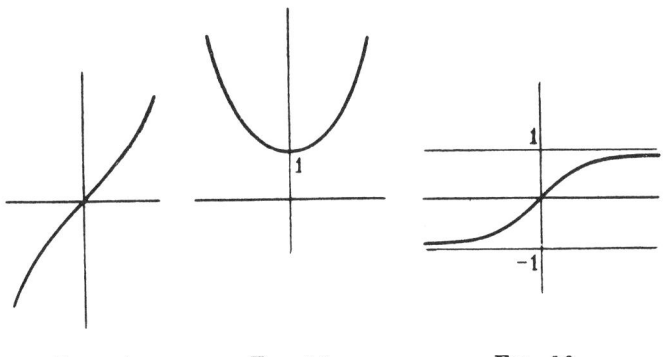

Fig. 14   Fig. 15   Fig. 16

(the value $y - \sqrt{y^2 + 1}$ being negative, we do not take it into account). In this way we have expressed x as a function of y. This is the required function inverse to $\sinh x$. We denote this function by the symbol ar sinh:

(20) $$\operatorname{ar sinh} x = \log\left(x + \sqrt{x^2 + 1}\right).$$

Similarly, we find the function inverse to $\cosh x$ ($x \geqslant 0$). This function is denoted by the symbol ar cosh. We find

(21) $$\operatorname{ar cosh} x = \log\left(x + \sqrt{x^2 - 1}\right), \quad x \geqslant 1.$$

Analogously, a function inverse to $\cosh x$ ($x \leqslant 0$) is $\log(x - \sqrt{x^2-1})$, $x \geqslant 1$.

Applying the formulae (16) and the formula on the derivative of the inverse function we find easily that

(22) $\quad \dfrac{d \operatorname{ar sinh} x}{dx} = \dfrac{1}{\sqrt{x^2+1}} = \dfrac{d}{dx} \log(x + \sqrt{x^2+1}),$

(23) $\quad \dfrac{d \operatorname{ar cosh} x}{dx} = \dfrac{1}{\pm \sqrt{x^2-1}} = \dfrac{d}{dx} \log(x \pm \sqrt{x^2-1}).$

This calculation may be performed also by differentiation of functions on the right sides of the formulae (22) and (23).

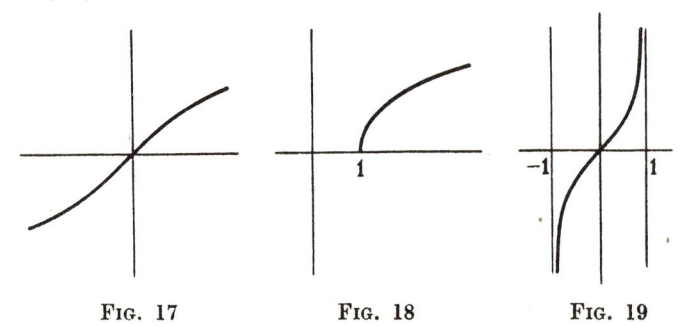

FIG. 17   FIG. 18   FIG. 19

We add that the following formula is found for the function inverse to $\tanh x$:

(24) $\quad \operatorname{ar tanh} x = \dfrac{1}{2} \log \dfrac{1+x}{1-x}, \quad |x| < 1.$

Hence

(25) $\quad \dfrac{d \operatorname{ar tanh} x}{dx} = \dfrac{1}{1-x^2} = \dfrac{d}{dx} \left(\dfrac{1}{2} \log \dfrac{1+x}{1-x} \right).$

Remark. The formulae on the differentiation of the fundamental functions with which we became acquainted in §§ 7.2 and 7.3 together with the formula for the differentiation of composite functions make it possible for us to differentiate an arbitrary elementary function (that is a function obtained from the fundamental functions by applying of an arbitrary number of compositions). A derivative of an elementary function is itself an elementary function.

Now let us consider the following function (non-elementary):

(26) $\quad f(x) = x^2 \cdot \sin\dfrac{1}{x} \quad$ for $\quad x \neq 0, \quad f(0) = 0$.

The general formulae for the differentiation of the elementary functions do not enable us to calculate $f'(0)$. We must calculate this derivative directly from the definition of a derivative.

We have

$$f'(0) = \lim_{h=0} \frac{f(h)-f(0)}{h} = \lim_{h=0} h \cdot \sin\frac{1}{h} = 0.$$

It is clear that for $x \neq 0$ the derivative $f'(x)$ may be calculated by means of general formulae concerning the differentiation of the elementary functions.

7.7. Geometrical interpretation of the sign of a derivative

Lagrange's theorem enables us to establish the relation between the sign of the derivative and the increase or decrease of the function.

1. *If the inequality $f'(x) > 0$ holds for any x belonging to an interval ab, then the function f is increasing in this interval.*

If $f'(x) < 0$ everywhere, then the function is decreasing.

According to the formula (11) we have for $h > 0$, $f(x+h) > f(x)$ or $f(x+h) < f(x)$, depending on whether we assume the derivative to be positive or negative everywhere. In the first case the function f increases and in the second one, decreases.

Remark. Assuming $f'(x) \geqslant 0$ or $f'(x) \leqslant 0$ everywhere, the function f is increasing in the wider sense or decreasing in the wider sense, respectively.

The following theorem is the converse to Theorem 1:

2. *If a function f is differentiable at a point c and increases (or decreases) in a certain interval surrounding this point, then $f'(c) \geq 0$ (or $f'(c) \leq 0$).*

Namely, if f increases, then for $h > 0$ we have $f(c+h) - f(c) > 0$ and so $\dfrac{f(c+h)-f(c)}{h} > 0$ and passing to the limit we obtain $f'(c) \geq 0$.

Similar arguments can be applied in the case in which the function f decreases.

We conclude from Theorem 1 that

3. *If $f'(c) > 0$, then the function f is increasing in a certain neighbourhood of the point c (assuming the derivative to be continuous at the point c).*

Similarly: *if $f'(c) < 0$, then the function is locally decreasing at the point c.*

Indeed, because of the continuity of the function f' the inequality $f'(c) > 0$ implies the existence of a number $\delta > 0$ such that there holds $f'(c+h) > 0$ for $|h| < \delta$. This means that the derivative $f'(x)$ is positive at any point of the interval $c-h < x < c+h$. Thus, according to Theorem 1 the function f is increasing in this interval.

Remark. Theorem 3 may be expressed in the following way: if $f'(c) \neq 0$, then (assuming the continuity of the derivative) the function f is locally one-to-one at the point c, i.e. it is one-to-one in a certain interval $c-\delta < x < c+\delta$ ($\delta > 0$). Thus, the function $y = f(x)$ possesses the inverse function $x = g(y)$ in this interval. As we already know, the derivative of this inverse function is equal to the inversion of the derivative of the function f.

By the above assumptions we see that for a given value of y the equation $y = f(x)$ possesses one and only one solution for x belonging to the interval $c-\delta$, $c+\delta$.

EXAMPLE. The function $y = \dfrac{e^x}{x}$ decreases for $0 < x < 1$ and increases for $x > 1$.
Indeed,
$$\frac{d}{dx}\left(\frac{e^x}{x}\right) = \frac{xe^x - e^x}{x^2} = \frac{e^x(x-1)}{x^2}.$$

This expression is <0 if $x < 1$ and >0 if $x > 1$.

Hence it follows that $\lim\limits_{x=\infty}\dfrac{e^x}{x} = \infty$. Indeed, $\lim\limits_{n=\infty}\dfrac{e^n}{n} = \infty$ (cf. § 3.5, 7) and so the function $\dfrac{e^x}{x}$, being unbounded and increasing, tends to ∞ as x tends to ∞ (we have obtained this result in another way in § 4.6, (β)).

More generally, we prove in a completely analogous way that the function $f(x) = e^x x^a$ ($x > 0$) is decreasing for $x < -a$ (and $a < 0$) and increasing for $x > -a$ and, since $\lim\limits_{n=\infty} e^n n^a = \infty$ (cf. § 3.5, 7),

(27) $$\lim_{x=\infty} e^x x^a = \infty.$$

7.8. Indeterminate expressions

Cauchy's theorem in § 7.5 makes it possible to calculate the limit of the following form:

(28) $$\lim_{x=a}\frac{f(x)}{g(x)}, \quad \text{where} \quad f(a) = 0 = g(a).$$

Expressions of this kind are called "indeterminate expressions of the type $\frac{0}{0}$".

1. *If the functions f and g are continuous in the closed interval $a \leqslant x \leqslant b$ and are differentiable inside this interval and if $f(a) = 0 = g(a)$, then*

(29) $$\lim_{x=a+0}\frac{f(x)}{g(x)} = \lim_{x=a+0}\frac{f'(x)}{g'(x)},$$

assuming the existence of the last limit.

III. DIFFERENTIAL CALCULUS

Let us write $x = a + h$. Then it has to be proved that

(30) $$\lim_{h=+0} \frac{f(a+h)}{g(a+h)} = \lim_{h=+0} \frac{f'(a+h)}{g'(a+h)}.$$

Now, the equations $f(a) = 0 = g(a)$ and the Cauchy formula give

$$\frac{f(a+h)}{g(a+h)} = \frac{f(a+h)-f(a)}{g(a+h)-g(a)} = \frac{f'(a+\Theta h)}{g'(a+\Theta h)}.$$

Since $\lim_{h=+0} \Theta h = +0$, we have

$$\lim_{h=+0} \frac{f'(a+\Theta h)}{g'(a+\Theta h)} = \lim_{h=+0} \frac{f'(a+h)}{g'(a+h)},$$

whence the formula (30) follows.

A similar theorem concerns the left-side limit.

When the derivatives f' and g' are continuous at the points a and $g'(a) \neq 0$, formula (29) implies immediately the following de l'Hospital ([1]) formula:

(31) $$\lim_{x=a} \frac{f(x)}{g(x)} = \frac{f'(a)}{g'(a)}.$$

A similar formula concerns the one-side limit and the one-side derivatives.

EXAMPLES. (α) To find $\lim_{x=0} \frac{\log(1+x)}{x}$ let us write

$$f(x) = \log(1+x) \quad \text{and} \quad g(x) = x.$$

Hence

$$f(0) = 0 = g(0), \quad f'(x) = \frac{1}{1+x}, \quad g'(x) = 1, \quad f'(0) = 1$$

and so

(32) $$\lim_{x=0} \frac{\log(1+x)}{x} = 1.$$

([1]) De l'Hospital, a French mathematician of the second half of the 17th century, known especially for popularizing the differential calculus.

(β) Evaluate $\lim_{x=0} \frac{e^x - 1}{x}$. Let us write $f(x) = e^x - 1$ and $g(x) = x$. Then we have $f(0) = 0 = g(0)$, $f'(x) = e^x$, $g'(x) = 1$. Hence

(33) $$\lim_{x=0} \frac{e^x - 1}{x} = \frac{f'(0)}{g'(0)} = 1.$$

Remarks. (α) If $g'(a) = 0$ but $f'(a) \neq 0$, then the formula (31) gives $\lim_{x=a} \frac{g(x)}{f(x)} = 0$. But if $f'(a) = 0 = g'(a)$, then the formula (31) cannot be applied. In this case the derivatives of higher orders have to be applied. (Cf. § 8.4.)

(β) The formula (29) may be applied also in the case when $a = \infty$. In other words, if $\lim_{x=\infty} f(x) = 0 = \lim_{x=\infty} g(x)$, then

(34) $$\lim_{x=\infty} \frac{f(x)}{g(x)} = \lim_{x=\infty} \frac{f'(x)}{g'(x)},$$

assuming the existence of the last limit.

We write $x = \frac{1}{t}$ and define an auxiliary function $F(t)$ by the conditions: $F(t) = f\left(\frac{1}{t}\right)$ for $t \neq 0$ and $F(0) = 0$. Similarly, let $G(t) = g\left(\frac{1}{t}\right)$ for $t \neq 0$ and $G(0) = 0$. The function F is continuous at any $t \neq 0$ as a composition of two continuous functions: $\frac{1}{t}$ and f. It is also right-side continuous at the point 0, since (cf. § 4, (18))

$$\lim_{t=+0} F(t) = \lim_{t=+0} f\left(\frac{1}{t}\right) = \lim_{x=\infty} f(x) = 0 = F(0).$$

Similarly, the function G is continuous. To be able to apply the formula (29), $\lim_{t=+0} \frac{F'(t)}{G'(t)}$ has to be calculated.

Now, we have

$$\frac{dF(t)}{dt} = \frac{df\left(\frac{1}{t}\right)}{dt} = \frac{df(x)}{dx} \cdot \frac{dx}{dt} = -\frac{1}{t^2} f'\left(\frac{1}{t}\right)$$

for $t \neq 0$. Similarly, $G(t) = -\frac{1}{t^2} g'\left(\frac{1}{t}\right)$. Thus,

$$\lim_{x=\infty} \frac{f(x)}{g(x)} = \lim_{t=+0} \frac{F(t)}{G(t)} = \lim_{t=+0} \frac{F'(t)}{G'(t)} = \lim_{t=+0} \frac{f'\left(\frac{1}{t}\right)}{g'\left(\frac{1}{t}\right)} = \lim_{x=\infty} \frac{f'(x)}{g'(x)}.$$

Besides indeterminate expressions of the form $\frac{0}{0}$, indeterminate expressions of the form $\frac{\infty}{\infty}$ are also considered. By such an expression we understand

$$\lim_{x=a} \frac{f(x)}{g(x)}, \quad \text{where} \quad \lim_{x=a} f(x) = \infty = \lim_{x=a} g(x).$$

Sometimes the calculation of indeterminate expressions of this form may be reduced to the form $\frac{0}{0}$. For, writing $F(x) = \frac{1}{f(x)}$ and $G(x) = \frac{1}{g(x)}$, we have

$$\lim_{x=a} \frac{f(x)}{g(x)} = \lim_{x=a} \frac{G(x)}{F(x)},$$

and this last indeterminate expression is of the form $\frac{0}{0}$.

Similarly, if $\lim_{x=a} f(x) = 0$ and $\lim_{x=a} g(x) = \infty$, then the indeterminate expression $\lim_{x=a} f(x) \cdot g(x)$ of the form $0 \cdot \infty$ may be reduced to the form $\frac{0}{0}$, for

$$\lim_{x=a} f(x) \cdot g(x) = \lim_{x=a} \frac{f(x)}{G(x)}.$$

7. DERIVATIVES OF THE FIRST ORDER

If $\lim_{x=a} f(x) = \infty = \lim_{x=a} g(x)$, then we reduce the indeterminate expression $\lim_{x=a}[f(x) - g(x)]$ of the form $\infty - \infty$ also to the form $\frac{0}{0}$, writing

$$\lim_{x=a}[f(x) - g(x)] = \lim_{x=a} \frac{G(x) - F(x)}{F(x) \cdot G(x)}.$$

It remains to consider indeterminate expressions of the form 0^0, ∞^0 and 1^∞. They may be reduced to the above considered by means of the relations

$$\lim_{x=a} f(x)^{g(x)} = \lim_{x=a} e^{g(x) \cdot \log f(x)} = e^{\lim_{x=a} g(x) \cdot \log f(x)}$$

EXAMPLES.

(γ) $\lim_{x=+0} (\cos x)^{\frac{1}{x}} = \lim_{x=+0} e^{\frac{1}{x}\log\cos x} = e^{\lim_{x\to+0}\frac{1}{x}\log\cos x} = 1$,

because

$$\lim_{x\to+0} \frac{\log\cos x}{x} = \lim_{x=+0} \frac{(\log\cos x)'}{x'} = \lim_{x=+0} \frac{-\tan x}{1} = 0.$$

(δ) Find the right-side derivative $f'_+(0)$ of the function f defined as follows: $f(x) = x^x$ when $x > 0$, $f(0) = 1$. We have

$$f'_+(0) = \lim_{h=+0} \frac{f(h) - f(0)}{h} = \lim_{h=+0} \frac{h^h - 1}{h}.$$

This is an expression of the form $\frac{0}{0}$, for $\lim_{x=+0} x^x = 1$ (§ 5.3). Applying formula (29) and taking into account the fact that $\frac{d}{dx}(x^x - 1) = x^x(\log x + 1)$ (cf. § 7.6, 3) we obtain

$$\lim_{x=+0} \frac{x}{x^x - 1} = \lim_{x=+0} \frac{1}{x^x(\log x + 1)} = -0,$$

whence $f'_+(0) = -\infty$.

(ε) The expression $\lim_{x=+0} (x \log x)$ is an indeterminate form of the type $0 \cdot \infty$. The proof that this limit is equal to 0 may be derived directly, without applying of the differential calculus (cf. § 4.6, (δ)).

7.9. The derivative of a limit

THEOREM. *If the equalities*

(35) $$\lim_{n=\infty} f_n(x) = f(x) \quad and \quad \lim_{n=\infty} f'_n(x) = g(x)$$

are satisfied in the interval $a \leqslant x \leqslant b$, *the functions* f'_n *being continuous and uniformly convergent to* g, *then*

(36) $$f'(x) = g(x), \quad i.e. \quad \frac{d}{dx} \lim_{n=\infty} f_n(x) = \lim_{n=\infty} \frac{df_n(x)}{dx}$$

Let c be a given point of the interval ab. Let us estimate the difference

(37) $$\frac{f_n(c+h) - f_n(c)}{h} - g(c) = f'_n(c + \Theta h) - g(c).$$

Let $\varepsilon > 0$ be given. Since the sequence $\{f'_n\}$ is uniformly convergent to g, there exists a k such that

(38) $$|f'_n(x) - g(x)| < \varepsilon$$

for $n > k$ and for any x belonging to the interval ab.

The function g being continuous as the limit of a uniformly convergent sequence of continuous functions (cf. § 6.1, 1), a number $\varepsilon > 0$ exists such that the inequality $|x - c| < \delta$ implies

(39) $$|g(x) - g(c)| < \varepsilon.$$

Thus, assuming $n > k$ and $|h| < \delta$ we obtain the following estimate:

$$|f'_n(c+\Theta h) - g(c)| \leqslant |f'_n(c+\Theta h) - g(c+\Theta h)| +$$
$$+ |g(c+\Theta h) - g(c)| < 2\varepsilon,$$

i.e., by (37),

$$\left| \frac{f_n(c+h) - f_n(c)}{h} - g(c) \right| < 2\varepsilon.$$

Since

$$\lim_{n=\infty} f_n(c+h) = f(c+h) \quad and \quad \lim_{n=\infty} f_n(c) = f(c),$$

we have
$$\left|\frac{f(c+h)-f(c)}{h}-g(c)\right|\leqslant 2\varepsilon,$$
if only $|h|<\delta$. Passing to the limit as h tends to 0 we therefore obtain equation (36).

Remarks. (α) Even the uniform convergence of a sequence of functions does not imply the convergence of the sequence of derivatives, as is shown by the following example:
$$f_n(x)=\frac{1}{n}\sin nx.$$

(β) The theorem remains true if we assume that the first of the equations (35) is satisfied at a given point c instead of assuming it to be satisfied in the whole interval ab. For, if the sequence of functions is convergent at a point c and if the sequence of the derivatives is uniformly convergent in the whole interval ab, then the sequence of functions is convergent in the whole interval ab.

Indeed, according to our assumptions, to a prescribed $\varepsilon>0$ there exists a number k such that the inequalities

(40) $\quad|f_n(c)-f_k(c)|<\varepsilon\quad$ and $\quad|f'_n(x)-f'_k(x)|<\varepsilon$

are satisfied for $n>k$ and $a\leqslant x\leqslant b$.

Let us write $F_n(x)=f_n(x)-f_k(x)$ and $x=c+h$. Then we have
$$F_n(x)=F_n(c)+h\cdot F'_n(c+\Theta h)$$
$$=f_n(c)-f_k(c)+h[f'_n(c+\Theta h)-f'_k(c+\Theta h)].$$

Hence, by (40), we have $|F_n(x)|<\varepsilon+(b-a)\varepsilon=\varepsilon(1+b-a)$. Thus the sequence $f_n(x)$ is convergent.

7.10. The derivative of a power series

We shall prove that

(41) $\quad\dfrac{d}{dx}(a_0+a_1x+a_2x^2+...)=a_1+2a_2x+3a_3x^2+...$

for any x lying within the interval of convergence of the series

(42) $$f(x) = a_0 + a_1 x + a_2 x^2 + \ldots$$

Moreover, the series (42) *and*

(43) $$g(x) = a_1 + 2a_2 x + 3a_3 x^2 + \ldots$$

have the same radius of convergence.

We shall first prove the second part of the theorem. Let r be the radius of convergence of the series (42) and let $0 < c < r$. We shall prove the series $g(c)$ to be convergent. Let $c < C < r$. Let us compare the terms of the series

(44) $$|a_1| + 2|a_2|c + \ldots + n|a_n|c^{n-1} + \ldots$$

with the corresponding terms of the convergent series (cf. § 6.3, Theorem 1):

(45) $$|a_1|C + |a_2|C^2 + \ldots + |a_n|C^n + \ldots$$

Since $\lim\limits_{n=\infty} \sqrt[n]{n} = 1$ (cf. § 3.5, 8) and $\dfrac{C}{c} > 1$, we have

$$\sqrt[n]{n} < \frac{C}{c}$$

and thus

$$n|a_n|c^{n-1} = \frac{|a_n|}{c}(\sqrt[n]{n}\,c)^n < \frac{|a_n|}{c}\left(\frac{C}{c} \cdot c\right)^n = \frac{1}{c}|a_n|C^n$$

for large n.

Hence we conclude that the series (44) is convergent and so the series $g(c)$ is absolutely convergent, too.

Conversely, if the series (43) is absolutely convergent, then the series (42) is also convergent. Namely,

$$|a_n x^n| = |x| \cdot |a_n x^{n-1}| \leqslant |x| \cdot |n a_n x^{n-1}|.$$

In this way we have proved that any point lying within the interval of convergence of the series (42) belongs to the interval of convergence of the series (43) and conversely: any point lying within the interval of convergence of the series (43) belongs to the interval of convergence of the series (42). Thus, both these intervals are equal.

Now, let us proceed to the proof of formula (41). Let x be an interior point of the interval of convergence of the series (42) and thus of the series (43), too. Let the interval ab be any closed interval surrounding this point and lying inside the interval of convergence. By Theorem 1 of § 6.3 the series (43) is uniformly convergent in the interval ab. This means that when we write

$$f_n(x) = a_0 + a_1 x + \ldots + a_n x^n,$$

the functions f'_n constitute a sequence uniformly convergent to the function g in the interval ab. Applying the theorem of the previous section we obtain the formula (41).

As is easily seen, formula (41) states that *a power series can be differentiated, term by term, like a polynomial.*

More generally, if by differentiating the series $\sum_{n=0}^{\infty} u_n(x)$ term by term we obtain the series $\sum_{n=0}^{\infty} u'_n(x)$ uniformly convergent and if, moreover, the functions $u'_n(x)$ are continuous, then

(46) $$\frac{d}{dx} \sum_{n=0}^{\infty} u_n(x) = \sum_{n=0}^{\infty} u'_n(x).$$

EXAMPLE. Differentiating the series

$$\frac{1}{1-x} = 1 + x + x^2 + \ldots + x^n + \ldots$$

we obtain for $|x| < 1$:

$$\frac{1}{(1-x)^2} = 1 + 2x + 3x^2 + \ldots + nx^{n-1} + \ldots$$

7.11. The expansion of the functions $\log(1+x)$ and $\arctan x$ in power series

Let us consider the known formula

(47) $\quad \dfrac{d\log(1+x)}{dx} = \dfrac{1}{1+x} = 1-x+x^2-x^3+\ldots \quad (|x|<1).$

It is obvious that the series appearing in this formula is the derivative of the series $S(x) = x - \dfrac{x^2}{2} + \dfrac{x^3}{3} - \dfrac{x^4}{4} + \dfrac{x^5}{5} - \ldots$
Hence the functions $\log(1+x)$ and $S(x)$ have the same derivative and so they differ only by a constant (cf. § 7.5, 4). Let us write $\log(1+x) = S(x) + C$. The constant C may be calculated by substituting $x = 0$; we obtain $\log(1+0) = 0 = S(0)$. Hence $C = \log(1+0) - S(0) = 0$.

So we have proved the functions $\log(1+x)$ and $S(x)$ to be identical in the interval $-1 < x < 1$. Speaking otherwise:

(48) $\quad \log(1+x) = x - \dfrac{x^2}{2} + \dfrac{x^3}{3} - \dfrac{x^4}{4} + \ldots \quad (|x|<1).$

Similarly, we deduce from the formula

$\dfrac{d\arctan x}{dx} = \dfrac{1}{1+x^2} = 1 - x^2 + x^4 - x^6 + \ldots \quad (|x|<1)$

that

$\arctan x = C + x - \dfrac{x^3}{3} + \dfrac{x^5}{5} - \ldots$

Substituting $x = 0$ and taking into account that $\arctan 0 = 0$, we obtain $C = 0$. Consequently,

(49) $\quad \arctan x = x - \dfrac{x^3}{3} + \dfrac{x^5}{5} - \dfrac{x^7}{7} + \ldots \quad (|x|<1).$

Thus, we have proved that the formula (48) holds for $|x| < 1$. We shall show that it holds also for $x = 1$. Indeed, for $x = 1$ the series on the right side of this formula, i.e. the series $S(1) = 1 - \tfrac{1}{2} + \tfrac{1}{3} - \ldots$, is convergent and thus by the Abel theorem (§ 6.3, 3) the function $S(x)$

is left-side continuous at the point 1. Hence
$$S(1) = \lim_{x=1-0} S(x) = \lim_{x=1-0} \log(1+x) = \log 2,$$
since the function $\log(1+x)$ is continuous at the point 1. Consequently,
(50) $\qquad \log 2 = 1 - \tfrac{1}{2} + \tfrac{1}{3} - \tfrac{1}{4} + \tfrac{1}{5} - \ldots$

Similarly, we deduce from formula (49) the Leibniz formula:
(51) $\qquad \dfrac{\pi}{4} = 1 - \dfrac{1}{3} + \dfrac{1}{5} - \dfrac{1}{7} + \dfrac{1}{9} - \ldots$

Denoting the right side of the equation (49) by $T(x)$ we have
$$T(1) = \lim_{x=1-0} T(x) = \lim_{x=1-0} \arctan x = \arctan 1 = \dfrac{\pi}{4},$$
whence formula (51) follows.

Remarks. (α) In practice the Leibniz formula is not convenient for the calculation of the number π; e.g. to calculate the first three decimals of the number $\pi/4$ it would be necessary to take the sum of the first 500 terms of the Leibniz expansion. Much more convenient is the following formula.

Let $\alpha = \arctan \tfrac{1}{2}$, $\beta = \arctan \tfrac{1}{3}$, i. e. $\tan \alpha = \tfrac{1}{2}$, $\tan \beta = \tfrac{1}{3}$. Since
$$\tan(\alpha+\beta) = \dfrac{\tan\alpha + \tan\beta}{1 - \tan\alpha \cdot \tan\beta} = \dfrac{\tfrac{1}{2}+\tfrac{1}{3}}{1-\tfrac{1}{2}\cdot\tfrac{1}{3}} = 1$$
and $1 = \arctan(\pi/4)$, so $\pi/4 = \alpha + \beta = \arctan \tfrac{1}{2} + \arctan \tfrac{1}{3}$. Hence, by formula (49), we obtain:
$$\dfrac{\pi}{4} = \dfrac{1}{2} - \dfrac{1}{3\cdot 8} + \dfrac{1}{5\cdot 32} - \ldots + \dfrac{1}{3} - \dfrac{1}{3\cdot 27} + \dfrac{1}{5\cdot 243} - \ldots$$

Other formulae giving even faster methods for the calculation of π are known.

(β) The formula (48) leads to the following inequality:
(52) $\qquad \dfrac{1}{x} \log(1+x) < 1 \quad \text{for} \quad 0 < x \leqslant 1.$

For, we have

(53) $\quad \dfrac{1}{x}\log(1+x) = 1 - \dfrac{x}{2} + \dfrac{x^2}{3} - \ldots \pm \dfrac{x^n}{n+1} \mp \ldots$

For $0 < x \leqslant 1$ this is an alternating series satisfying the conditions

$$1 > \dfrac{x^2}{2} > \dfrac{x^3}{3} > \ldots \quad \text{and} \quad \lim_{n=\infty} \dfrac{x^n}{n} = 0;$$

thus, according to the theorem § 3.3, the sum of this series is less than its first term, i. e. the inequality (52) holds.

Similarly we prove that

(54) $\quad \left(\dfrac{1}{x} + \dfrac{1}{2}\right)\log(1+x) > 1 \quad \text{for} \quad 0 < x \leqslant 1$,

since adding to the series (53) the series (48) divided by 2, we obtain

$$\left(\dfrac{1}{x} + \dfrac{1}{2}\right)\log(1+x)$$
$$= 1 + x^2\left(\dfrac{1}{3} - \dfrac{1}{4}\right) - x^3\left(\dfrac{1}{4} - \dfrac{1}{6}\right) + \ldots \pm x^n\left(\dfrac{1}{n+1} - \dfrac{1}{2n}\right) \mp \ldots$$

It is easily seen that by omitting the first term we obtain an alternating series satisfying the assumptions of Theorem 1 of § 3.3. Hence we conclude that the sum of this series is positive. Thus, formula (54) holds.

7.12*. Asymptotes

By applying the differential calculus it is possible to find the asymptotes of a given curve $y = f(x)$. The straight line $Y = aX + b$ is called an *asymptote* of the curve $y = f(x)$, if

$$a = \lim_{x=\infty} f'(x) \quad \text{and} \quad b = \lim_{x=\infty} [f(x) - ax];$$

hence it is seen that the direction of an asymptote is a limiting direction to which the direction of the tangent to the curve at the point $[x, f(x)]$ tends, as x tends to ∞;

moreover, the distance between this point and the asymptote tends to 0, because this distance is equal to $|[f(x) - ax - b]\cos a|$, where a is the angle between the asymptote and the X-axis.

Similarly we consider asymptotes for x tending to $-\infty$.

Finally, the straight line $X = c$ (parallel to the Y-axis) is called an asymptote of the curve $y = f(x)$, $x < c$, if $\lim_{x=c-0} f'(x) = \pm\infty = \lim_{x=c-0} f(x)$. Similarly we define the asymptotes to the curve $y = f(x)$, $x > c$.

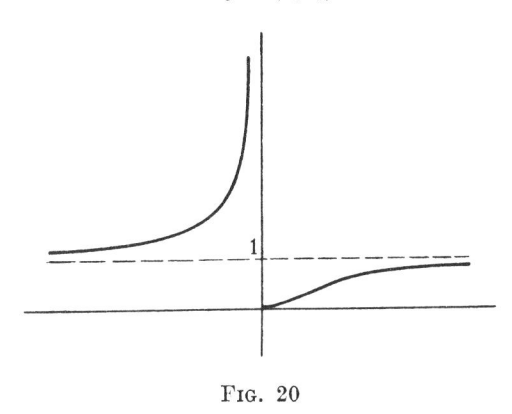

Fig. 20

EXAMPLES. As it is easily seen, the asymptotes of the hyperbola $y = 1/x$ are the axes X and Y.

To find the asymptotes of the curve $y = e^{-1/x}$, we calculate $y' = \dfrac{e^{-1/x}}{x^2}$. Hence $\lim_{x=\infty} y = 1$ and $\lim_{x=\infty} y' = 0$. Thus, $a = 0$ and $b = 1$, and the straight line $Y = 1$ is an asymptote. A second asymptote is the Y-axis, because $\lim_{x=-0} y = \infty$ and $\lim_{x=-0} y' = \infty$.

7.13*. The concept of a differential

A rigorous mathematical sense may be given to the concept of a differential in the following way. By the *differential* of a function $y = f(x)$ at a point x with regard to the increment h we understand the product $f'(x)h$. We write

(55) $\quad df(x) = f'(x)h \quad$ or, more briefly, $\quad dy = f'(x)h$.

The symbol $df(x)$ or dy, convenient in applications for its simplicity, takes into account neither the dependence of the differential on the variable x with respect to which the differentiation is performed, nor the dependence on the increment h. To satisfy these requirements we should have to write e.g. $d_x(y, h)$. Moreover, if we should like to point out that the differential is considered at a point x_0, we should have to write $[d_x(f(x), h)]_{x=x_0}$ by analogy with the notation for the derivative at a point x_0.

By formula (55), $dx = h$, for substituting $f(x) = x$ we have $f'(x) = 1$.

Thus we may substitute in the formula (55) $h = dx$. We obtain $dy = f'(x)dx$. Dividing both sides of this equation by dx we have then

$$f'(x) = \frac{dy}{dx},$$

where the right side is the quotient of two differentials (which may be considered as an explanation of the Leibniz notation for the derivatives). Speaking more strictly, we have $f'(x) = \dfrac{d_x(y, h)}{d_x(x, h)}$ for every $h \neq 0$. The formulae of the differential calculus may be written in the differential notation as is seen by the following examples:

$$de^x = e^x dx, \quad d\sin x = \cos x\, dx, \quad dc = 0.$$

Formula 4 of § 7.4 for the derivative of a sum leads to an analogous formula for the differential of a sum:

(56) $\quad\quad\quad\quad d(y+z) = dy + dz$.

Indeed, let $y = f(x)$, $z = g(x)$. We have

$$d(y+z) = (y+z)'h = y'h + z'h = dy + dz.$$

7. DERIVATIVES OF THE FIRST ORDER

Analogously, we prove that

(57) $$d(yz) = y\,dz + z\,dy.$$

Now we shall give the formula for the differential of a composition of functions.

Let $y = f(x)$ and $z = g(y)$. Then

(58) $$d_x z = g'(y)d_x y.$$

Namely, $d_x z = \dfrac{dgf(x)}{dx}dx = g'(y)f'(x)dx = g'(y)d_x y$, according to (55).

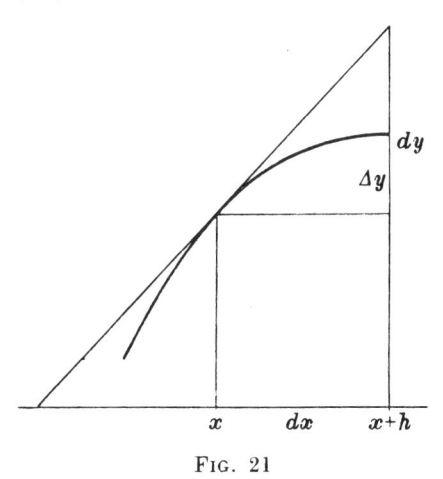

FIG. 21

Let us note that by (55) we have also $d_y z = g'(y)d_y y$. Comparing this equation with equation (58) we see that if z is a function of the variable y, then the formula for the differential z remains formally unchanged after representing y as a function of a new variable x (however, the index at d showing with respect to which variable one has to differentiate has to be obviously omitted).

The geometrical interpretation of the differential is immediately seen in Fig. 21.

The following theorem holds: *The difference between the increment of the function $\Delta y = f(x+h) - f(x)$ and the* *differential dy, divided by the differential dx tends to 0 together with this differential, i.e.*

(59) $$\lim_{dx=0} \frac{\Delta y - dy}{dx} = 0.$$

Namely,

$$\frac{\Delta y - dy}{dx} = \frac{f(x+h) - f(x) - f'(x)h}{h} = \frac{f(x+h) - f(x)}{h} - f'(x),$$

and

$$\lim_{h=0} \frac{f(x+h) - f(x)}{h} = f'(x).$$

In practice (e.g. in physical applications) the above theorem often makes it possible to replace the differential dy by the increment Δy.

Exercises on § 7

1. Differentiate the following functions:

1) $\dfrac{x^2 + 2x + 4}{x^4 - 1}$, 2) $\sqrt{2x - x^5}$, 3) $\dfrac{x}{\sqrt{c + x^3}}$,

4) $\sec x$, 5) $\operatorname{cosec} x$, 6) $\cot x$,

7) $\operatorname{arc sec} x$, 8) $\operatorname{arc cosec} x$, 9) $\operatorname{arc cotan} x$,

10) $\log \sin x$, 11) $\log \tan x$, 12) $\arccos(1-x)$,

13) $\arctan \dfrac{5x}{1-x^2}$, 14) $\log \sinh x$, 15) $\operatorname{ar tanh} \sqrt[3]{x}$.

2. Prove that if a function f is differentiable, then its derivative f' possesses the Darboux property (i.e. passes from one value to another through all the intermediate values).

(First prove that if $f'(a) < 0 < f'(b)$, then a point c exists in the interval ab such that $f'(c) = 0$. The general case when $f'(a) < A < f'(b)$ may be easily reduced to the previous one considering the function $f(x) - Ax$.)

3. Given the series $f(x) = \sum_{n=1}^{\infty} \left(\frac{x^n}{n} - \frac{x^{n+1}}{n+1} \right)$, find $f'(x)$ and the sum of the series of the derivatives. Compare the result obtained with the theorem in § 7.9.

4. Prove that if a function f is differentiable at a point x, then
$$f'(x) = \lim_{h=0} \frac{f(x+h)-f(x-h)}{2h}.$$

Show by an example that the converse relation does not hold: the above limit (i.e. the so called *generalized derivative*) may exist although the function is not differentiable at the point x.

5. Give an example of a function having a generalized derivative at a point of discontinuity.

6. Give an example of an even function (i. e. such that $f(x) = f(-x)$) for which $f'(0) = 0$, although the function does not possess an extremum (even an improper) at the point 0.

7. Prove that if a continuous function possesses maxima at the points a and b, then it possesses also a minimum at a certain point between a and b.

8. Let a function f be continuous in an interval $a \leqslant x \leqslant b$ and differentiable inside this interval. If the right-side limit $\lim_{x=a+0} f'(x)$ exists, then the right-side derivative of the function f exists, too, and they are both equal: $f'_+(a) = \lim_{x=a+0} f'(x)$.

9. Prove that among all rectangles with the same perimeter the square has the greatest area.

10. Prove that among all triangles with a constant perimeter and with a constant base the isosceles triangle has the greatest area.

11. Prove that among all triangles with a constant perimeter the equilateral triangle has the greatest area.

12. Prove that if the derivative f' is strictly monotone, then for any x_0 the curve $y = f(x)$ lies on one side of the tangent at this point.

Hence conclude that $e^x > 1+x$ for $x \neq 0$.

13. Prove that $\left(\dfrac{1}{x}+\dfrac{1}{3}\right)\log(1+x) < 1$ for $0 < x \leqslant 1$.

14. Prove that $\dfrac{x}{x+1} \leqslant \log(1+x) \leqslant x$ for $x > -1$.

15. Expand $\log\sqrt{\dfrac{1+x}{1-x}}$ in a power series.

16. Given two points P and Q lying on the plane above the X-axis, find a point R on the X-axis such that the sum of the segments PR and RQ is the shortest possible. Prove that the (acute) angles between these segments and the X-axis are equal.

(Interpret this result in the theory of optics: PRQ = the path of the ray, the X-axis = the mirror).

17. Evaluate $\lim\limits_{x=0}\dfrac{\tan x}{\tan ax}$, $\lim\limits_{x=\frac{\pi}{2}-0}\left(\dfrac{\pi}{2}-x\right)\tan x$, $\lim\limits_{x=\infty}\sqrt[x]{x}$.

18. Prove the existence of the limit (called the *Euler constant*):

$$C = \lim_{n=\infty}\left(1+\dfrac{1}{2}+\dfrac{1}{3}+\ldots+\dfrac{1}{n}-\log n\right)$$

(prove that the considered sequence is decreasing and bounded, applying the inequality from Exercise 14).

19. Prove that if the differentiable functions f and g satisfy the inequality $f'(x)g(x) \neq g'(x)f(x)$ for every x, then between any two roots of the equation $f(x) = 0$ there exists a root of the equation $g(x) = 0$. (Consider the auxiliary function $\dfrac{f(x)}{g(x)}$.)

§ 8. DERIVATIVES OF HIGHER ORDERS

8.1. Definition and examples

The derivative of the derivative of a function f is called the *second derivative* of this function. Similarly, the third derivative is the derivative of the second de-

8. DERIVATIVES OF HIGHER ORDERS

rivative. Generally, the n-th derivative is the derivative of the $(n-1)$-th derivative. We denote the derivatives of higher orders by expressions of the kind:

$$f', f'', f''', \ldots, f^{(n)}, \ldots \quad \text{or} \quad \frac{dy}{dx}, \frac{d^2y}{dx^2}, \frac{d^3y}{dx^3}, \ldots, \frac{d^ny}{dx^n}, \ldots$$

So we have

$$\frac{d^ny}{dx^n} = \frac{d}{dx}\left(\frac{d^{n-1}y}{dx^{n-1}}\right).$$

E.g. let $f(x) = x^3$. We have $f'(x) = 3x^2$, $f''(x) = 6x$, $f'''(x) = 6$ and $f^{(n)}(x) = 0$ for $n > 3$. More generally,

(1) $\quad \dfrac{d^k(x^n)}{dx^k} = n(n-1) \cdot \ldots \cdot (n-k+1) x^{n-k} \quad \text{for} \quad k \leqslant n,$

$\dfrac{d^n(x^n)}{dx^n} = n!$

If a is a real number (but not a positive integer), then we have

(2) $\quad \dfrac{d^k}{dx^k}(1+x)^a = a(a-1) \cdot \ldots \cdot (a-k+1)(1+x)^{a-k}$

for an arbitrary positive integer k and for an arbitrary $x > -1$.

The function e^x has the property that its derivatives of all orders are identical with itself.

Taking $f(x) = \sin x$ we have

(3) $\quad f'(x) = \cos x, \quad f''(x) = -\sin x,$

$\quad\quad f'''(x) = -\cos x, \quad f^{IV}(x) = \sin x.$

Here the equation $f^{(n)}(x) = f^{(n+4)}(x)$ holds. The same equation is satisfied by the function $f(x) = \cos x$.

Let us note by the way that it is convenient to consider the derivative of order 0 as equal to the function $f^{(0)}(x) = f(x)$. The above given equation holds also for $n = 0$.

The formula (1) makes it possible to calculate easily the k-th derivative of a polynomial of degree n: $f(x) =$

$= a_0 + a_1 x + \ldots + a_n x^n$. Substituting $x = 0$ in $f^{(k)}(x)$ (for $k = 0, 1, \ldots, n$) we find:

(4) $\quad a_0 = f(0), \quad a_1 = f'(0), \quad a_2 = \frac{1}{2} f''(0), \quad \ldots, \quad a_n = \frac{1}{n!} f^{(n)}(0).$

Hence we obtain the formula

(5) $\quad f(x) = f(0) + xf'(0) + \frac{x^2}{2!} f''(0) + \ldots + \frac{x^n}{n!} f^{(n)}(0)$

which we shall generalize below.

As an example of the calculation of derivatives of higher orders let us mention the formula for the second derivative of the inverse function.

As we know, $\dfrac{dx}{dy} = 1 : \dfrac{dy}{dx}$. Hence

$$\frac{d^2 x}{dy^2} = \frac{d\left(1 : \dfrac{dy}{dx}\right)}{dx} \cdot \frac{dx}{dy} = - \frac{\dfrac{d^2 y}{dx^2}}{\left(\dfrac{dy}{dx}\right)^2} \cdot \frac{dx}{dy},$$

i.e.

$$\frac{d^2 x}{dy^2} = - \frac{d^2 y}{dx^2} : \left(\frac{dy}{dx}\right)^3.$$

In applications in physics the second derivative is often of importance. In particular, the acceleration of a particle as a derivative of the velocity with respect to the time is the second derivative of the distance covered with respect to the time, i.e. $\dfrac{d^2 s}{dt^2}$. Hence the force acting on the particle is equal to $m \dfrac{d^2 s}{dt^2}$, where m denotes the mass of the particle.

8.2*. Differentials of higher order

We define the second differential of a function f as the differential of the differential of the function f. Speaking more strictly, we define it as the differential

8. DERIVATIVES OF HIGHER ORDERS

with respect to the increment h of the differential of the function f with respect to the increment h; the variable with respect to which the differentiation is performed is x. Then, writing $y = f(x)$ and denoting the second differential by d^2y we have

$$d^2y = d(dy) = d(y'h) = \frac{d}{dx}(y'h)h = y''h^2 = y''(dx)^2.$$

Hence we see that the second derivative of a function f is the quotient of its second differential over the square of the differential dx (which explains the notation of the second derivative).

Generally, the n-th differential $d^n y$ is defined as the n-th iteration of the differential, i.e. $d^n y = d(d^{n-1}y)$ (with respect to the same increment h).

Then we have

(6) $$d^n y = y^{(n)}(dx)^n.$$

We prove this by induction. Formula (6) holds for $n = 1$ (and as we have seen for $n = 2$, too). Assuming this formula for n, it has to be proved for $n+1$. Now,

$$d^{n+1}y = d(d^n y) = d[y^{(n)}(dx)^n] = \frac{d}{dx}[y^{(n)}(dx)^n]dx$$
$$= y^{(n+1)}(dx)^{n+1}$$

EXAMPLE. The derivatives of higher orders of the inverse function (cf. the previous section) may be calculated by means of differentials of higher orders as follows. Let the function $x = g(y)$ be inverse to $y = f(x)$. The differentials dy, d^2y, d^3y etc. will be understood as $d_y y$, $d_y^2 y$, $d_y^3 y$ etc. Evidently, $d^2y = 0$, $d^3y = 0$, ...

Since $y'x' = 1$, we have $dy = y'x'dy$. Differentiating with respect to y, we obtain

$$0 = d^2y = \left(\frac{dy'}{dx}\frac{dx}{dy}x' + \frac{dx'}{dy}y'\right)(dy)^2 = [y''(x')^2 + y'x''](dy)^2.$$

Similarly,
$$0 = d^3y = [y'''(x')^3 + 3y''x'x'' + y'x'''](dy)^3.$$

In this way we obtain equations which make it possible for us to calculate the derivatives x', x'', x''', ..., successively:
$$y'x' = 1,$$
$$y''(x')^2 + y'x'' = 0,$$
$$y'''(x')^3 + 3y''x'x'' + y'x''' = 0,$$
. . . .

8.3. Arithmetical operations

Let $y = f(x)$, $z = g(x)$. We verify easily by induction that
$$(7) \qquad (y+z)^{(n)} = y^{(n)} + z^{(n)}$$
and
$$(8) \qquad (y-z)^{(n)} = y^{(n)} - z^{(n)}.$$

For the n-th derivative of a product, the following Leibniz formula holds:
$$(9) \quad (yz)^{(n)} = y^{(n)}z + \binom{n}{1} y^{(n-1)}z' + \binom{n}{2} y^{(n-2)}z'' + \ldots + yz^{(n)}$$
$$= \sum_{k=0}^{n} \binom{n}{k} y^{(n-k)} z^{(k)}.$$

We shall prove this formula by induction. For $n = 1$, we know this formula as the formula for the derivative of a product. Let us assume this formula to hold for n and let us differentiate it. We obtain

$$(yz)^{(n+1)} = \sum_{k=0}^{n} \left\{ \binom{n}{k} y^{(n-k+1)} z^{(k)} + \binom{n}{k} y^{(n-k)} z^{(k+1)} \right\}$$
$$= y^{(n+1)}z + \sum_{k=1}^{n} \left\{ y^{(n-k+1)} z^{(k)} \left[\binom{n}{k} + \binom{n}{k-1} \right] \right\} + yz^{(n+1)}$$
$$= \sum_{k=0}^{n+1} \binom{n+1}{k} y^{(n+1-k)} z^{(k)},$$

since

$$\binom{n}{k} + \binom{n}{k-1} = \binom{n+1}{k}$$

(cf. § 1, (4)).

EXAMPLE. To find the n-th derivative of the function xe^x, let us write $y = e^x$ and $z = x$. As is easily seen, only the two first terms do not vanish in the formula (9). Hence

$$\frac{d^n}{dx^n}(xe^x) = e^x(x+n).$$

8.4. Taylor formula [1]

The Lagrange formula (§ 7, (11)) is the "first approximation" of the following Taylor formula:

Let us assume the function f to be n times differentiable in an interval $a \leqslant x \leqslant b$. Let us write $h = b - a$. Then $f(b)$ may be represented in the following form:

$$(10) \quad f(b) = f(a) + \frac{b-a}{1!}f'(a) + \frac{(b-a)^2}{2!}f''(a) + \ldots$$
$$+ \frac{(b-a)^{n-1}}{(n-1)!}f^{(n-1)}(a) + R_n,$$

where

$$(11) \quad R_n = \frac{h^n}{n!}f^{(n)}(a + \Theta h) = \frac{h^n(1-\Theta')^{n-1}}{(n-1)!}f^{(n)}(a + \Theta' h),$$

where Θ and Θ' are suitably chosen numbers satisfying the inequalities $0 < \Theta < 1$ and $0 < \Theta' < 1$.

Proof. According to equality (10), we have

$$(12) \quad R_n = f(b) - f(a) - \frac{b-a}{1!}f'(a) - \frac{(b-a)^2}{2!}f''(a) - \ldots$$
$$- \frac{(b-a)^{n-1}}{(n-1)!}f^{(n-1)}(a).$$

---

[1] Brook Taylor and Colin Maclaurin (cited in the following), respectively English and Scottish mathematicians of the first half of the 18th century.

Let us denote by $g_n(x)$ an auxiliary function obtained from R_n by replacing a by x, i.e.

(13) $\quad g_n(x) = f(b) - f(x) - \dfrac{b-x}{1!} f'(x) -$

$$- \dfrac{(b-x)^2}{2!} f''(x) - \ldots - \dfrac{(b-x)^{n-1}}{(n-1)!} f^{(n-1)}(x).$$

Differentiating we obtain

$$g_n'(x) = -f'(x) + \left[f'(x) - \dfrac{b-x}{1!} f''(x) \right] +$$

$$+ \left[2 \dfrac{b-x}{2!} f''(x) - \dfrac{(b-x)^2}{2!} f'''(x) \right] + \ldots +$$

$$+ \left[(n-1) \dfrac{(b-x)^{n-2}}{(n-1)!} f^{(n-1)}(x) - \dfrac{(b-x)^{n-1}}{(n-1)!} f^{(n)}(x) \right].$$

Thus, $g_n'(x) = -\dfrac{(b-x)^{n-1}}{(n-1)!} f^{(n)}(x)$. At the same time $g_n(b) = 0$, $g_n(a) = R_n$.

Applying the mean-value theorem, we obtain

$$\dfrac{g_n(b) - g_n(a)}{b-a} = g_n'(a + \Theta' h),$$

i.e.

$$\dfrac{-R_n}{h} = -\dfrac{(b-a-\Theta' h)^{n-1}}{(n-1)!} f^{(n)}(a + \Theta' h)$$

$$= -\dfrac{h^{n-1}(1-\Theta')^{n-1}}{(n-1)!} f^{(n)}(a + \Theta' h),$$

whence

$$R_n = \dfrac{h^n (1-\Theta')^{n-1}}{(n-1)!} f^{(n)}(a + \Theta' h),$$

according to the second part of formula (11); this is the so called *Cauchy-form of the remainder*.

To prove the first part of formula (11), i.e. the Lagrange formula for the remainder, we shall apply to the functions $g_n(x)$ and $u_n(x) = (b-x)^n$ the Cauchy theorem (§ 7.5, 5). We obtain

$$\frac{g_n(b)-g_n(a)}{u_n(b)-u_n(a)} = \frac{g_n'(a+\Theta h)}{u_n'(a+\Theta h)}.$$

Taking into account the equations $u_n(b) = 0$, $u_n(a) = h^n$ and $u_n'(x) = -n(b-x)^{n-1}$ we conclude that

$$\frac{-R_n}{-h^n} = -\frac{(b-a-\Theta h)^{n-1}}{(n-1)!} f^{(n)}(a+\Theta h) - \frac{1}{-n(b-a-\Theta h)^{n-1}},$$

whence the remainder R_n is obtained in the *Lagrange-form*.

Thus, the Taylor theorem is proved completely.

Substituting $b = x$ and $a = 0$, we obtain the *Maclaurin formula*:

(14) $\quad f(x) = f(0) + \dfrac{x}{1!}f'(0) + \ldots + \dfrac{x^{n-1}}{(n-1)!} f^{(n-1)}(0) + R_n$,

where

(15) $\qquad R_n = \dfrac{x^n}{n!} f^{(n)}(\Theta x) = \dfrac{x^n(1-\Theta')^{n-1}}{(n-1)!} f^{(n)}(\Theta' x).$

The above formula is satisfied assuming the function f to be n times differentiable in the closed interval $0x$ or $x0$, according as to whether $x > 0$ or $x < 0$. (As is easily seen from the proof, the assumption of the n times differentiability of the function may be replaced by a weaker assumption, namely, it is sufficient to assume that the $(n-1)$-th derivative is continuous in the whole interval and that the n-th derivative exists inside the interval; these are the assumptions which we have made for $n = 1$ in Rolle theorem).

EXAMPLES AND APPLICATIONS. (α) Let us apply the Maclaurin formula substituting $f(x) = e^x$ and $n = 2$ in formula (14). Since

$$f'(x) = e^x = f''(x) \quad \text{and} \quad f(0) = 1 = f'(0),$$

we have

$$e^x = 1 + x + \frac{x^2}{2} e^{\Theta x}.$$

Hence we conclude that for any x the inequality
(16) $$e^x \geqslant 1+x$$
holds, because $x^2 \geqslant 0$ and $e^{\Theta x} > 0$.

Inequality (16) possesses a clear geometrical content on the graphs of the functions $y = e^x$ and $y = 1+x$.

(β) The Taylor formula makes it possible to sharpen the de l'Hospital formula (§ 7, (31)) as follows: if the functions f and g possess continuous n-th derivatives and if

$$f(a) = 0, \ f'(a) = 0, \ \ldots, \ f^{(n-1)}(a) = 0,$$
$$g(a) = 0, \ g'(a) = 0, \ \ldots, \ g^{(n-1)}(a) = 0, \quad \text{but} \quad g^{(n)}(a) \neq 0,$$

then

(17) $$\lim_{x=a} \frac{f(x)}{g(x)} = \frac{f^{(n)}(a)}{g^{(n)}(a)}.$$

Indeed, according to our assumptions and the Taylor formula (10) (with b replaced by x), we obtain

$$f(x) = \frac{(x-a)^n}{n!} f^{(n)}[a + \Theta(x-a)]$$

and

$$g(x) = \frac{(x-a)^n}{n!} g^{(n)}[a + \Theta'(x-a)].$$

Thus,

$$\frac{f(x)}{g(x)} = \frac{f^{(n)}[a+\Theta(x-a)]}{g^{(n)}[a+\Theta'(x-a)]}.$$

Passing to the limit for x tending to a and basing on the continuity of the functions $f^{(n)}$ and $g^{(n)}$, we obtain formula (17).

For example to calculate $\lim\limits_{x=0} \dfrac{x-\sin x}{x \sin x}$, we write $f(x) = x-\sin x$ and $g(x) = x \sin x$. Thus we have

$f(0) = 0 = g(0), \quad f'(x) = 1-\cos x,$
$\qquad\qquad\qquad\qquad g'(x) = \sin x + x \cos x,$
$f'(0) = 0 = g'(0), \quad f''(x) = \sin x,$
$\qquad\qquad\qquad\qquad g''(x) = 2\cos x - x \sin x,$
$\qquad f''(0) = 0, \quad g''(0) = 2.$

Applying formula (17) (for $n = 2$) we obtain

(18) $$\lim_{x=0} \frac{x - \sin x}{x \sin x} = \frac{0}{2} = 0.$$

Similarly we find

(19) $$\lim_{x=0} \left(\cot x - \frac{1}{x} \right) = 0.$$

For, $\cot x - \frac{1}{x} = \frac{x \cos x - \sin x}{x \sin x}$ and writing

$$f(x) = x \cos x - \sin x \quad \text{and} \quad g(x) = x \sin x$$

we obtain the required formula applying a calculation completely similar to the previous one.

(γ) Applying the Maclaurin formula to the function $f(x) = \log(1 + x)$ and taking into account that

$$f'(x) = \frac{1}{1+x}, \quad f''(x) = -\frac{1}{(1+x)^2}, \quad f(0) = 0, \quad f'(0) = 1$$

we get

$$\log(1+x) = x - \frac{x^2}{2(1+\Theta x)^2}$$

for $x > -1$.

In particular, we have, for $x = \frac{1}{n}$,

$$0 < \frac{1}{n} - \log \frac{n+1}{n} = \frac{1}{2n^2 \left(1 + \frac{1}{n} \Theta_n\right)^2} < \frac{1}{2n^2}.$$

Since the series $\sum_{n=1}^{\infty} \frac{1}{n^2}$ is convergent (cf. § 3, (14)), the series $\sum_{n=1}^{\infty} \left(\frac{1}{n} - \log \frac{n+1}{n} \right)$ is convergent, too. We denote the sum of the last series by γ. Thus we have

$$\gamma = \lim_{n=\infty} \sum_{k=1}^{n} \left(\frac{1}{k} - \log \frac{k+1}{k} \right)$$
$$= \lim_{n=\infty} \left(1 + \frac{1}{2} + \ldots + \frac{1}{n} - \log \frac{2}{1} \cdot \frac{3}{2} \cdot \ldots \cdot \frac{n+1}{n} \right)$$
$$= \lim_{n=\infty} \left[1 + \frac{1}{2} + \ldots + \frac{1}{n} - \log(n+1) \right].$$

But $\lim\limits_{n=\infty} [\log(n+1) - \log n] = \lim\limits_{n=\infty} \log \dfrac{n+1}{n} = 0$; consequently, we obtain (cf. § 7, Exercise 18):

(20) $$\gamma = \lim_{n=\infty} \left(1 + \dfrac{1}{2} + \ldots + \dfrac{1}{n} - \log n\right).$$

Approximately, the Euler constant $\gamma = 0{,}5772\ldots$ ([1]).

8.5. Expansions in power series

We conclude immediately from the formula (12) that if $\lim\limits_{n=\infty} R_n = 0$, then

(21) $$f(b) = \sum_{n=0}^{\infty} \dfrac{(b-a)^n}{n!} f^{(n)}(a).$$

In particular, if we have $\lim\limits_{n=\infty} R_n = 0$ in the Maclaurin formula, then the function $f(x)$ may be expanded in a power series:

(22) $$f(x) = f(0) + \dfrac{x}{1!} f'(0) + \dfrac{x^2}{2!} f''(0) + \ldots = \sum_{n=0}^{\infty} \dfrac{x^n}{n!} f^{(n)}(0).$$

Before proceeding to applications of the above theorem we note that if all derivatives $f^{(n)}$ are uniformly bounded in the interval $0x$, i.e. if a number M exists such that the inequality $M > |f^{(n)}(\Theta x)|$ holds for each n and for any Θ satisfying the condition $0 < \Theta < 1$, then $f(x)$ has the expansion (22) in a (Maclaurin) power series.

For, $|R_n| = \left|\dfrac{x^n}{n!} f^{(n)}(\Theta x)\right| \leqslant \left|\dfrac{x^n}{n!}\right| \cdot M$, and since $\lim\limits_{n=\infty} \dfrac{x^n}{n!} = 0$, we have also $\lim\limits_{n=\infty} R_n = 0$.

---

[1] Till now it is not known whether γ is a rational number or an irrational number.

Leonhard Euler (1707-1783), a great Swiss mathematician; mathematics owes him, besides many new results, a systematic elaboration of the mathematical analysis of those times.

8. DERIVATIVES OF HIGHER ORDERS

Remarks. (α) The convergence of the series appearing on the right side of formula (22) is not a sufficient condition of equation (22), i.e. of the expansibility of the function $f(x)$ in a power series.

For example the function $f(x) = e^{-\frac{1}{x^2}}$ (for $x \neq 0$), $f(0) = 0$, cannot be expanded in a power series although its Maclaurin series is convergent; we have here $f^{(n)}(0) = 0$ for each n.

(β) Formula (4) may be generalized to a general power series. For, if $f(x) = a_0 + a_1 x + a_2 x^2 + \dots$, then $a_0 = f(0)$, $a_1 = \frac{1}{1!} f'(0)$, and generally,

$$(23) \qquad a_n = \frac{1}{n!} f^{(n)}(0).$$

In other words, if a function possesses an expansion in a power series, then it possesses only one such expansion, namely, the Maclaurin expansion (formula (22)).

To prove this statement, we shall first prove by induction that the assumption

$$f(x) = \sum_{n=0}^{\infty} a_n x^n$$

implies

$$f^{(k)}(x) = \sum_{n=k}^{\infty} n(n-1) \cdot \ldots \cdot (n-k+1) a_n x^{n-k}.$$

Substituting in this formula $x = 0$, we obtain equation (23).

APPLICATIONS. 1. $e^x = 1 + \frac{x}{1!} + \frac{x^2}{2!} + \dots = \sum_{n=0}^{\infty} \frac{x^n}{n!}$.

To prove this formula, we note that writing $f(x) = e^x$ we have $f^{(n)}(x) = e^x$ and $f^{(n)}(0) = 1$. Moreover, the derivatives of all orders are uniformly bounded; for, if $0 \leqslant x$, then $f^{(n)}(\Theta x) \leqslant e^x$ and if $0 > x$, then $f^{(n)}(\Theta x) < 1$.

In particular: 1'. $e = 1 + \frac{1}{1!} + \frac{1}{2!} + \dots$

2. $\sin x = \dfrac{x}{1!} - \dfrac{x^3}{3!} + \dfrac{x^5}{5!} - \ldots$

Applying formula (3) of § 8.1 we conclude immediately that the derivatives of all orders of the function sin are uniformly bounded (by the number 1) in the set of all real numbers. Moreover, $f^{(n)}(0) = \sin 0 = 0$ for $n = 0, 2, 4, \ldots$ and $f'(0) = 1$, $f'''(0) = -1$, $f^{V}(0) = 1, \ldots$ Similarly, we prove that

3. $\cos x = 1 - \dfrac{x^2}{2!} + \dfrac{x^4}{4!} - \ldots$

We shall deduce now the Newton binomial formula for arbitrary real exponents (which are not positive integers) and for $|x| < 1$:

4. $(1+x)^a = 1 + ax + \dfrac{a(a-1)}{2!} x^2 + \ldots$

$\qquad = \sum_{n=0}^{\infty} \dfrac{a(a-1) \cdot \ldots \cdot (a-n+1)}{n!} x^n$.

We shall consider the cases of x positive and of x negative, separately.

1° $x > 0$. Since

$$f^{(n)}(x) = a(a-1) \cdot \ldots \cdot (a-n+1)(1+x)^{a-n}$$

(cf. (2)), the remainder in the Lagrange-form is of the following form:

$$R_n(x) = \dfrac{a(a-1) \cdot \ldots \cdot (a-n+1)}{n!} x^n (1 + \Theta_n x)^{a-n} \; (^1).$$

Since (cf. § 3.4, 6) $\lim\limits_{n=\infty} \dfrac{a(a-1) \cdot \ldots \cdot (a-n+1)}{n!} x^n = 0$, for $|x| < 1$, to prove that $\lim\limits_{n=\infty} R_n(x) = 0$ it suffices to show that for a given x the sequence $(1+\Theta_n x)^{a-n}$, $n = 0, 1, 2, \ldots$, is bounded. Now, according to the inequality $x > 0$ we have $1 < 1 + \Theta_n x < 1 + x$. Thus, $1 \leqslant (1 + \Theta_n x)^a \leqslant (1+x)^a$

(^{1}) We add the index n to Θ in order to point out that Θ depends on n.

or $(1+x)^a \leqslant (1+\Theta_n x)^a \leqslant 1$ according as whether $a \geqslant 0$ or $a \leqslant 0$. Moreover, $(1+\Theta_n x)^{-n} < 1$.

Hence the considered sequence is bounded.

$2°$ $x < 0$. Writing the remainder in the Cauchy-form, we have

$$R_n(x) = \frac{a(a-1)\cdot\ldots\cdot(a-n+1)}{(n-1)!} x^n (1-\Theta'_n)^{n-1}(1+\Theta'_n x)^{a-n}.$$

Just as before, we have

$$\lim_{n=\infty} \frac{(a-1)(a-2)\cdot\ldots\cdot(a-n+1)}{(n-1)!} x^{n-1} = 0;$$

it is to be proved that the sequence $(1-\Theta'_n)^{n-1}(1+\Theta'_n x)^{a-n}$ is bounded, i. e. that the sequences $\left(\dfrac{1-\Theta'_n}{1+\Theta'_n x}\right)^{n-1}$ and $(1+\Theta'_n x)^{a-1}$ are bounded. For this purpose we write $y = -x$. Then $y > 0$ and $\Theta'_n > \Theta'_n y$, whence $1-\Theta'_n < 1-\Theta'_n y < 1$. Thus, $\left(\dfrac{1-\Theta'_n}{1-\Theta'_n y}\right)^{n-1} < 1$.

On the other hand, $1-y < 1-\Theta'_n y < 1$ and so

$$(1-y)^{a-1} \leqslant (1-\Theta'_n y)^{a-1} \leqslant 1$$

or $\quad 1 \leqslant (1-\Theta'_n y)^{a-1} \leqslant (1-y)^{a-1}$

according as whether $a-1 \geqslant 0$ or $a-1 \leqslant 0$. Thus, both sequences are bounded. Hence $\lim\limits_{n=\infty} R_n(x) = 0$.

EXAMPLES. Substituting $a = \tfrac{1}{2}$ we obtain

$$\sqrt{1+x} = 1 + \frac{1}{2}x - \frac{1}{2\cdot 4}x^2 + \frac{1\cdot 3}{2\cdot 4\cdot 6}x^3 - \ldots$$

For $a = -\tfrac{1}{2}$, we have

$$\frac{1}{\sqrt{1+x}} = 1 - \frac{1}{2}x + \frac{1\cdot 3}{2\cdot 4}x^2 - \frac{1\cdot 3\cdot 5}{2\cdot 4\cdot 6}x^3 + \ldots$$

Hence

$$\frac{1}{\sqrt{1-x^2}} = 1 + \frac{1}{2}x^2 + \frac{1\cdot 3}{2\cdot 4}x^4 + \frac{1\cdot 3\cdot 5}{2\cdot 4\cdot 6}x^6 + \ldots$$

The above formula makes it possible to apply to the expansion of the function $\arcsin x$ in a power series the same method which we have applied expanding the functions $\log x$ and $\arctan x$. Namely, as is easily seen, the series appearing in this formula is the derivative of the series

$$S(x) = x + \frac{1}{2} \cdot \frac{x^3}{3} + \frac{1 \cdot 3}{2 \cdot 4} \cdot \frac{x^5}{5} + \frac{1 \cdot 3 \cdot 5}{2 \cdot 4 \cdot 6} \cdot \frac{x^7}{7} + \ldots$$

But

$$\frac{d}{dx} \arcsin x = \frac{1}{\sqrt{1-x^2}}; \quad \text{hence} \quad \arcsin x = S(x) + C.$$

At the same time $\arcsin 0 = 0 = S(0)$ and so $C = 0$, i.e. $\arcsin x = S(x)$. In this way we have the following formula:

5. $\arcsin x = x + \frac{1}{2} \cdot \frac{x^3}{3} + \frac{1 \cdot 3}{2 \cdot 4} \cdot \frac{x^5}{5} + \frac{1 \cdot 3 \cdot 5}{2 \cdot 4 \cdot 6} \cdot \frac{x^7}{7} + \ldots,\ |x| < 1.$

We note that substituting $x = \frac{1}{2}$ and taking into account the fact that $\sin \frac{1}{6} \pi = \frac{1}{2}$ we obtain:

$$\frac{\pi}{6} = \frac{1}{2} + \frac{1}{2} \cdot \frac{1}{3 \cdot 2^3} + \frac{1 \cdot 3}{2 \cdot 4} \cdot \frac{1}{5 \cdot 2^5} + \frac{1 \cdot 3 \cdot 5}{2 \cdot 4 \cdot 6} \cdot \frac{1}{7 \cdot 2^7} + \ldots$$

8.6. A criterion for extrema

We have proved in § 7.4, 2 that if a (differentiable) function possesses an extremum at a point x, then $f'(x) = 0$. However, the converse theorem is not true: the equation $f'(x) = 0$ does not imply the existence of an extremum at the point x; the example of the function $f(x) = x^3$ shows this. The investigation of the derivatives of higher orders leads to the following more precise criterion for the existence of an extremum.

Let us assume that $f'(c) = 0$, $f''(c) = 0$, ..., $f^{(n-1)}(c) = 0$ and $f^{(n)}(c) \neq 0$. Then, if n is an even number, the function f has a proper extremum at the point c, namely a maximum, if $f^{(n)}(c) < 0$, and a minimum, if $f^{(n)}(c) > 0$. But if n is an odd number, then the function does not possess an extremum at the point c. (We assume the continuity of the n-th derivative at the point c).

Indeed, applying the Taylor formula we have

$$(24) \quad f(c+h) = f(c) + \frac{h}{1!}f'(c) + \ldots + \frac{h^{n-1}}{(n-1)!}f^{(n-1)}(c) + \\ + \frac{h^n}{n!}f^{(n)}(c+\Theta h),$$

whence according to the assumptions of our theorem we obtain

$$(25) \quad f(c+h) - f(c) = \frac{h^n}{n!}f^{(n)}(c+\Theta h).$$

Let us assume n to be an even number and let $f^{(n)}(c) < 0$. Because of the continuity of the function $f^{(n)}$ at the point c we know there exists a $\delta > 0$ such that the inequality $|x-c| < \delta$ implies $f^{(n)}(x) < 0$. Since $|\Theta h| < \delta$ for $|h| < \delta$, so $f^{(n)}(c+\Theta h) < 0$. Hence we conclude by (25) that if $0 < |h| < \delta$, then $f(c+h) - f(c) < 0$ (because $h^n > 0$ for even n). This means that the function f possesses a maximum at the point c.

It is proved by analogous arguments, that if n is an even number and if $f^{(n)}(c) > 0$, then $f(c+h) - f(c) > 0$ and thus the function has a minimum at the point c.

Now, let us assume that n is an odd number. Let $f^{(n)}(c) < 0$ (in the case $f^{(n)}(c) > 0$ the arguments are analogous). As before we choose a number $\delta > 0$ in such a way that $f^{(n)}(c+\Theta h) < 0$ for $|h| < \delta$. Then for $0 < h < \delta$ we have the inequality $f(c+h) - f(c) < 0$, i. e. $f(c) > f(c+h)$ and for $-\delta < h < 0$, we have the inequality $f(c+h) - f(c) > 0$, showing that $f(c) < f(c+h)$. Thus, there is neither a maximum nor a minimum at the point c.

EXAMPLE. The function $f(x) = x^n$ possesses a minimum at the point 0, if n is an even number. If n is odd, then there is no extremum at this point.

Indeed, the first derivative of the function x^n not vanishing at the point $x = 0$ is the n-th derivative (cf. (1)) and $f^{(n)}(x) = n!$

Remark. The previous theorem does not determine for all functions (arbitrarily many times differentiable) whether an extremum exists at a given point (e.g. for the function $e^{-\frac{1}{x^2}}$ considered in § 8.4 all derivatives of which vanish at the point 0). However, it always permits us to answer this question for functions having a Taylor expansion (formula (21)) in a neighbourhood of a given point a. For, it follows from formula (21) that if the function f is not constant, then not all derivatives of this function vanish at the point a.

8.7. Geometrical interpretation of the second derivative. Points of inflexion

As we have seen in § 7.7, 3, if $f'(c) > 0$, then the function increases in a neighbourhood of the point c. Hence if $f''(c) > 0$, then the function f' increases; thus, assuming $f'(c) > 0$, the function f increases faster and faster.

1. *If $f''(c) > 0$, then in a certain neighbourhood of the point c the curve $y = f(x)$ lies above the tangent of this curve at the point $[c, f(c)]$ (thus, it is directed with the convexity downwards).*

Analogously, if $f''(c) < 0$, then this curve lies locally below the tangent (thus, it is directed with the convexity upwards).

In both cases we assume the continuity of the second derivative.

Indeed, let us estimate the difference between the differences quotient and the derivative

$$\varphi(h) = \frac{f(c+h) - f(c)}{h} - f'(c).$$

8. DERIVATIVES OF HIGHER ORDERS

According to the Taylor formula, we have

$$f(c+h)-f(c)-hf'(c) = \frac{h^2}{2}f''(c+\Theta h).$$

Hence, assuming $f''(c) > 0$, for sufficiently small h we also have $f''(c+\Theta h) > 0$, from which it follows that $f(c+h)-f(c)-hf'(c) > 0$. Interpreting the differences quotient as the tangent of the angle between the secant and the X-axis we conclude that for $h > 0$ this tangent

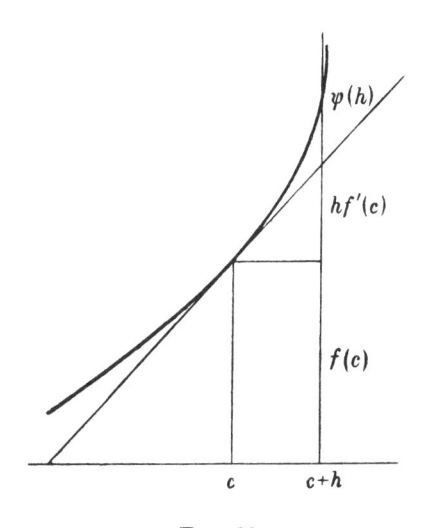

Fig. 22

is greater than $f'(c)$, i.e. greater than the tangent of the angle between the tangent and the X-axis. This means that the arc of the curve lies above the tangent.

The proof of the second part of the theorem is completely analogous.

EXAMPLE. The parabola $y = x^2$ lies, in the neighbourhood of any point, above the tangent at this point. The second derivative is equal to 2.

A curve $y = f(x)$ is said to possess a *point of inflexion* at the point c, if for sufficiently small increments h ($|h| < \delta$) the arc of the curve lies for positive increments

on the other side of the tangent to the curve at the point $[c, f(c)]$ from what it does for negative increments. In other words: if a number $\delta > 0$ exists such that the expression
$$\psi(h) = f(c+h) - f(c) - hf'(c)$$
is positive for $0 < h < \delta$ and negative for $0 > h > -\delta$ or conversely: negative for $0 < h < \delta$ and positive for $0 > h > -\delta$.

EXAMPLE. The sinusoidal curve $y = \sin x$ has a point of inflexion at the point 0. Indeed, $\sin h - \sin 0 - h \cos 0 = \sin h - h$. This difference is negative for $h > 0$ and is positive for $h < 0$.

Similarly, the cubic parabola $y = x^3$ possesses a point of inflexion at the point 0.

It follows from theorem 1 that if $f''(c) \neq 0$, then the curve lies (locally) on one side of the tangent. Hence the point c is not a point of inflexion. In other words:

2. *If a point c is a point of inflexion of a curve $y = f(x)$, then $f''(c) = 0$.*

This theorem cannot be reversed: the second derivative may vanish at a point c although the point c is not a point of inflexion (just as the first derivative may vanish although there is no extremum at the given point); e.g. the curve $y = x^4$ has a minimum at the point 0 and so this point is not a point of inflexion of this function although the second derivative $12x^2$ vanishes for $x = 0$.

To get a more precise theorem, further derivatives have to be considered. This leads to the following theorem:

3. *If $f''(c) = 0, f'''(c) = 0, \ldots, f^{(n-1)}(c) = 0$ but $f^{(n)}(c) \neq 0$ then in the case of an odd n the curve $y = f(x)$ possesses a point of inflexion at the point c; in the case of an even n, c is not a point of inflexion.* (We assume the continuity of the n-th derivative at the point c).

Indeed, applying the Taylor formula (24) we obtain,
$$\psi(h) = f(c+h) - f(c) - hf'(c) = \frac{h^n}{n!} f^{(n)}(c + \Theta h).$$

Let us assume n to be an odd number and let $f^{(n)}(c) > 0$ (the arguments in the case $f^{(n)}(c) < 0$ are analogous). Then for sufficiently small increments h we have $f^{(n)}(c + \Theta h) > 0$. Hence we have $\psi(h) > 0$ for $h > 0$ and $\psi(h) < 0$ for $h < 0$. Thus, the point c is a point of inflexion of the curve $y = f(x)$.

On the other hand, if n is an even number, then $h^n > 0$ independently of the sign of h, whence $\psi(h)$ has the same sign as $f^{(n)}(c)$. Hence the curve lies on one side of the tangent to the curve at the point $[c, f(c)]$. Thus, the point c is not a point of inflexion.

Remark. As in the case of the extrema, the question whether c is a point of inflexion may always be answered (according to Theorem 3) for functions expansible in Taylor series in a neighbourhood of the point c. At the same time the points of inflexion of the curve $y = f(x)$ are extrema of the curve $y = f'(x)$.

Exercises on § 8

1. Prove the (Halphen) formula:
$$\frac{d^n}{dx^n}(x^{n-1} \cdot e^{\frac{1}{x}}) = (-1)^n x^{-n-1} \cdot e^{\frac{1}{x}}.$$

2. Find the n-th derivative of the functions: $\dfrac{\log x}{x}$, $e^x \cos x$.

3. Evaluate:
$$\lim_{x \to 0} \frac{e^x - e^{-x} - 2x}{x - \sin x}, \quad \lim_{x \to 0} \frac{x - \sin x}{x^3}, \quad \lim_{x \to 0}\left(\frac{1}{x^2} - \cot^2 x\right).$$

4. Expand the functions $\sinh x$ and $\cosh x$ in power series.

5. Sketch the graphs of the functions x^x, $e^{\frac{1}{x}}$, $\log \sin x$, showing their extrema, points of inflexion, asymptotes, points of one-side discontinuity or points of both-sides discontinuity (if any exist).

6. Prove that if an n times differentiable function vanishes at $n+1$ different points of a given interval, then there exists in this interval a point at which the n-th derivative vanishes.

7. Let $f(x) = \dfrac{1}{x^n} e^{-\frac{1}{x^2}}$ for $x \neq 0$ and $f(0) = 0$. Prove that $f'(0) = 0$. Deduce as a corollary that the function g defined by the conditions

$$g(x) = e^{-\frac{1}{x^2}} \quad \text{for} \quad x \neq 0 \quad \text{and} \quad g(0) = 0$$

possesses at the point 0 derivatives of all orders equal to 0 (cf. § 8.5, Remark (α)).

8. Prove that $e^n < (2n+1)\dfrac{n^n}{n!}$. Deduce that

$$\lim_{n=\infty} \frac{1}{n} \sqrt[n]{n!} = \frac{1}{e}.$$

CHAPTER IV

INTEGRAL CALCULUS

§ 9. INDEFINITE INTEGRALS

9.1. Definition

A function F is called a *primitive function* of a function f defined in an open interval (finite or infinite), if $F'(x) = f(x)$ for every x.

For example the function $\sin x$ is a primitive function of the function $\cos x$. Any function of the form $\sin x + C$, where C is a constant, is also a primitive function of the function $\cos x$.

If a function f is defined in a closed interval $a \leqslant x \leqslant b$, then the function F is called its primitive function, if $F'(x) = f(x)$ for $a < x < b$, $F'_+(a) = f(a)$ and $F'_-(b) = f(b)$.

1. *If two functions F and G are primitive functions of a function f in an interval ab (open or closed), then these two functions differ by a constant.*

Indeed, if $F'(x) = G'(x)$, then, according to Theorem 4 of § 7.5, there exists a constant C such that $G(x) = F(x) + C$ for every x.

Conversely, a function obtained by adding a constant to a primitive function of a function f is also a primitive function of the function f. Thus the expression $F(x) + C$ is the general form of a primitive function of the function f. We indicate this expression by the symbol $\int f(x)dx$ ("the integral $f(x)dx$") and we call it the *indefinite integral of the function f*. So we have

1) $\quad \int f(x)dx = F(x) + C, \quad$ where $\quad F'(x) = f(x)$,

(2) $$\frac{d}{dx}\int f(x)\,dx = f(x),$$

(3) $$\int \frac{dF(x)}{dx}\,dx = F(x) + C.$$

The evaluation of the indefinite integral of a function f, i.e. the calculation of the primitive function of a function f is called the *integration of the function f*. So integration is an inverse operation to differentiation.

It follows from the definition of the indefinite integral, that any formula for the derivative of a function automatically gives a formula for the integral of another function (namely, the derived function). For example from the formula $\frac{d\sin x}{dx} = \cos x$, we obtain $\int \cos x\,dx = \sin x + C$. In general, however, the problem of the calculation of the integral of a continuous function, which we do not know to be a derivative of a certain function, is more difficult than the problem of differentiation. As we have seen in § 7, the differentiation of functions which are compositions of elementary functions does not lead out of their domain; yet an analogous theorem for indefinite integrals would not be true. It is known only that every continuous function possesses an indefinite integral (cf. § 9.2). However, this theorem gives us no practical procedure of evaluating the indefinite integral of a given continuous function.

EXAMPLES. Applying well-known formulae from the differential calculus we obtain immediately the following formulae:

(4) $\int 0\,dx = C$,

(5) $\int a\,dx = ax + C$,

(6) $\int x^n\,dx = \frac{1}{n+1}x^{n+1} + C$,

(7) $\int \cos x\,dx = \sin x + C$,

(8) $\int \sin x\,dx = -\cos x + C$,

(9) $\int \dfrac{dx}{\cos^2 x} = \tan x + C$,

(10) $\int \dfrac{dx}{x} = \log|x| + C$,

(11) $\int e^x\,dx = e^x + C$

(12) $\int \dfrac{dx}{\sqrt{1-x^2}} = \arcsin x + C$,

(13) $\int \dfrac{dx}{1+x^2} = \arctan x + C$,

(14) $\int x^a\,dx = \dfrac{1}{a+1} x^{a+1} + C$, if $a \neq -1$ and $x > 0$.

Remark. If the domain of x for which the equation $F'(x) = f(x)$ is satisfied is not an interval (finite or infinite), then it cannot be stated that the expression $F(x) + C$ gives all primitive functions of the function f in this domain of arguments.

For example, $\log|x| + C$ gives all primitive functions of the function $\dfrac{1}{x}$ in each of the two domains $x < 0$ and $x > 0$, separately, but not in the whole domain of x real and different from 0. Namely, the function $G(x)$ defined as $\log|x|$ for $x < 0$ and as $\log|x| + 1$ for $x \neq 0$ is a primitive function of the function $\dfrac{1}{x}$ for all $x \neq 0$ although it is not given by the formula $\log|x| + C$.

Let us complete now Theorem 1 as follows:

2. *Let a point x_0 be given inside an interval ab and let an arbitrary real number y_0 be given. If a function f possesses a primitive function in the interval ab, then it possesses one and only one primitive function F such that $F(x_0) = y_0$.*

Let $P(x)$ be an arbitrary primitive function of the function $f(x)$ in an interval ab. Let us write $F(x) = P(x) - P(x_0) + y_0$. Then we have $F'(x) = P'(x) = f(x)$ and $F(x_0) = y_0$. Hence the function F satisfies the conditions of the theorem.

Moreover, it is the only function satisfying these conditions, since any other primitive function of the function f is of form $F(x)+C$, where $C \neq 0$, and this implies $F(x_0)+C = y_0+C \neq y_0$.

Geometrically, this theorem means that given any point on the plane with the abscissa belonging to the interval ab, there exists an integral curve (i.e. the graph of a primitive function) passing through this point. The integral curves being parallel one to another, only one integral curve of a given function f may pass through a given point on the plane.

9.2. The integral of the limit. Integrability of continuous functions

We have proved in § 7.9 that given a sequence of functions $F_n(x)$, $a \leqslant x \leqslant b$, such that the derivatives are continuous and uniformly convergent in the interval ab to a function $g(x)$ and that the sequence $F_n(c)$ is convergent for a certain point c belonging to the interval ab, the sequence $F_n(x)$ is convergent for every x belonging to this interval; moreover, writing $F(x) = \lim\limits_{n=\infty} F_n(x)$, we have $F'(x) = g(x)$.

Hence, the following lemma follows:

1. *If the functions $f_n(x)$ are continuous and uniformly convergent in an interval ab to a function $f(x)$ and if they possess primitive functions, then the function $f(x)$ possesses also a primitive function.*

The primitive functions $F_n(x)$ may be chosen in such a way (according to § 9.1, 2) that the equation $F_n(c) = 0$ holds for a certain point c of the interval ab, for each value of n. Then

$$\int f(x)\,dx = \lim_{n=\infty} F_n(x) + C,$$

i.e. the integral of the limit equals to the limit of the integrals.

Indeed, to obtain our lemma it is sufficient to substitute in the above formulation f_n in place of F'_n and f in place of g.

2. THEOREM. *Every function continuous in an interval ab possesses a primitive function in this interval.*

By the theorem proved in § 6.4, every function continuous in the interval ab is a limit of a uniformly convergent sequence of polygonal functions. Hence, according to the lemma it remains only to prove that any polygonal function f possesses a primitive function.

According to the definition of a polygonal function there exists a system of $n+1$ points $a_0 < a_1 < ... < a_n$, where $a_0 = a$, $a_n = b$, and two systems of numbers $c_1, c_2, ..., c_n$ and $d_1, d_2, ..., d_n$ such that

$$f(x) = c_k x + d_k \quad \text{for} \quad a_{k-1} \leqslant x \leqslant a_k \quad (k = 1, 2, ..., n).$$

Let us write

$$F_k(x) = \tfrac{1}{2} c_k x^2 + d_k x + e_k \quad \text{for} \quad a_{k-1} \leqslant x \leqslant a_k,$$

where $e_1 = 0$ and

$$e_{k+1} = \tfrac{1}{2} c_k a_k^2 + d_k a_k + e_k - (\tfrac{1}{2} c_{k+1} a_k^2 + d_{k+1} a_k)$$

for $k \geqslant 1$.

Then $F_k(a_k) = F_{k+1}(a_k)$, whence the collection of the functions $F_1, F_2, ..., F_n$ defines one function equal to each of the functions of this collection in the suitable interval, respectively.

Differentiating the function F_k we obtain immediately $F'(x) = f(x)$, i.e. the function F is primitive with respect to the function f.

9.3. General formulae for integration

Let us assume the functions f and g to be continuous. Then the following formulae hold:

1. $\quad \int [f(x) + g(x)] dx = \int f(x) dx + \int g(x) dx$,

since $\dfrac{d}{dx} \left(\int f(x) dx + \int g(x) dx \right) = f(x) + g(x)$.

2. $\int af(x)\,dx = a\int f(x)\,dx$,

since $\dfrac{d}{dx}\left(a\int f(x)\,dx\right) = a\dfrac{d}{dx}\int f(x)\,dx = af(x)$.

EXAMPLE. The integral of a polynomial:

$$\int (a_0 + a_1 x + \ldots + a_n x^n)\,dx = C + a_0 x + \frac{a_1}{2}x^2 + \frac{a_2}{3}x^3 + \ldots + \frac{a_n}{n+1}x^{n+1}.$$

3. *The formula for the integration by parts*:

$$\int f(x)g'(x)\,dx = f(x)g(x) - \int f'(x)g(x)\,dx$$

(if the functions f' and g' are continuous).

We write this formula more briefly, taking $y = f(x)$ and $z = g(x)$ as follows:

$$\int y\frac{dz}{dx}\,dx = yz - \int z\frac{dy}{dx}\,dx.$$

To prove this formula, let us differentiate its right side. We obtain: $yz' + y'z - zy' = yz'$. This is the integrated function of the left side of the formula. Thus the formula is proved.

A special case of formula 3 is the following formula obtained from the formula 3 by substitution $z = x$:

3. $\int y\,dx = yx - \int x\dfrac{dy}{dx}\,dx$.

EXAMPLE. $\int \log x\,dx = x\log x - \int x\dfrac{1}{x}\,dx = x\log x - x + C$.

4. *The formula for integration by substitution (i.e. the formula for the change of the variable)*:

$$\int g[f(x)]\frac{df(x)}{dx}\,dx = \int g(y)\,dy,$$

where on the right side it has to be substituted $y = f(x)$ after evaluation of the integral.

We rewrite this formula more briefly, writing $y = f(x)$ and $z = g(y)$, as follows:

$$\int z \frac{dy}{dx} dx = \int z\, dy.$$

Let us write $G(y) = \int g(y)\, dy$. To prove the formula 4 it has to be shown that the derivative of the function $G[f(x)]$ is equal to the integrated function on the left side of this formula. Now,

$$\frac{dG[f(x)]}{dx} = \left(\frac{dG(y)}{dy}\right)_{y=f(x)} \cdot \frac{df(x)}{dx} = g[f(x)] \frac{df(x)}{dx}. \quad \text{Q. E. D.}$$

As is seen, from the point of view of the calculation we may reduce by the differential. The formula 3 may be written also in the following form:

$$\int y\, dz = yz - \int z\, dy.$$

EXAMPLES. (α) $\int g(ax)\, dx = \dfrac{1}{a} \int g(y)\, dy$, where $y = ax$.

Namely, substituting $f(x) = ax$ in the formula 4, we have

$$\frac{1}{a} \int g(ax)\, a\, dx = \frac{1}{a} \int g(y)\, dy.$$

(β) Similarly we prove that

$$\int f(x+a)\, dx = \int f(y)\, dy, \quad \text{where } y = x+a.$$

(γ) [1] To evaluate the integral $\int \dfrac{x\, dx}{1+x^2}$ we substitute $y = x^2$. We obtain:

$$\int \frac{x\, dx}{1+x^2} = \int \frac{1}{1+x^2} \cdot \frac{1}{2} \cdot \frac{d(x^2)}{dx}\, dx$$
$$= \int \frac{1}{2} \cdot \frac{dy}{1+y} = \frac{1}{2} \log(1+y) = \frac{1}{2} \log(1+x^2).$$

---

[1] To simplify the calculation, we omit the constant C in the example (γ) and in the further examples.

Since—as we said—by the integration the differential $df(x)$ may be substituted in place of $f'(x)dx$, i.e. in the given case, $d(x^2)$ in place of $2x\,dx$, the above calculation may be performed in a little shorter way:

$$\int \frac{x\,dx}{1+x^2} = \int \frac{1}{2} \cdot \frac{dx^2}{1+x^2} = \frac{1}{2}\log(1+x^2).$$

(δ) We have for $n > 1$

$$\int \frac{x\,dx}{(1+x^2)^n} = \frac{1}{2}\int \frac{dx^2}{(1+x^2)^n} = -\frac{1}{2(n-1)} \cdot \frac{1}{(1+x^2)^{n-1}}.$$

(ε) $\quad \displaystyle\int \tan x\,dx = \int \frac{\sin x}{\cos x}dx = -\int \frac{d\cos x}{\cos x} = -\log|\cos x|.$

The formulae (γ) and (ε) are easily obtained from the following general formula:

(ζ) $\quad \displaystyle\int \frac{1}{y} \cdot \frac{dy}{dx}\,dx = \int \frac{dy}{y} = \log|y| + C.$

Substituting $y = 1+x^2$ in formula (δ) we deduce this formula easily from the following general formula:

(η) $\quad \displaystyle\int y^a \frac{dy}{dx}\,dx = \int y^a\,dy = \frac{1}{a+1}y^{a+1}+C \quad (a \ne -1).$

As is seen, the method of integration in the above examples consists in finding a function $y = f(x)$ such that the given integral will be transformed in an integral with respect to y (which is easier to calculate). This makes it possible to replace the calculation of the integral on the left side of the formula 4 by the calculation of the integral on the right side of this formula. Sometimes, the reverse procedure, i.e. the replacing of the right side by the left side is more suitable; to calculate the integral $\int g(y)dy$ we look for a strictly monotone function $y = f(x)$ such that the integral $\int g[f(x)]\dfrac{df(x)}{dx}dx$ may be easily calculated. Denoting by $x = h(y)$ the function inverse to $y = f(x)$ and by $F(x)$ and $G(y)$ the integrals

on the left side and on the right side of the formula 4, respectively, we have then
$$F(x) = G[f(x)], \quad \text{i.e.} \quad G(y) = F[h(y)].$$
In other words,
$$\int g(y)\,dy = \int g[f(x)]\frac{df(x)}{dx}\,dx,$$
where in the integral on the right side has to be substituted $x = h(y)$.

EXAMPLE. Let us substitute in the following integral $x = \sin t$, $-\frac{\pi}{2} < t < \frac{\pi}{2}$:
$$\int \sqrt{1-x^2}\,dx = \int \cos t \frac{d\sin t}{dt}\,dt = \int \cos^2 t\,dt,$$
but $\cos^2 t = \frac{1}{2}(1+\cos 2t)$, whence $\int \cos^2 t\,dt = \frac{1}{2}t + \frac{1}{4}\sin 2t$. Since $\sin 2t = 2\sin t \cos t = 2\sin t\sqrt{1-\sin^2 t}$, we obtain hence:
$$\int \sqrt{1-x^2}\,dx = \tfrac{1}{2}t + \tfrac{1}{2}\sin t\sqrt{1-\sin^2 t}$$
$$= \tfrac{1}{2}\arcsin x + \tfrac{1}{2}x\sqrt{1-x^2}.$$

Remark. We verify the correctness of the integration by differentiating the function obtained as the result of the integration. After the differentiation we should obtain the integrated function.

5. *Recurrence methods for calculation of integrals.* The recurrence method for calculation of the integral $\int f_n(x)\,dx$ consists in calculation of the integral for $n = 1$ (or for $n = 0$) and in reducing the n-th integral to the $n-1$-th one (or an even earlier one).

EXAMPLES. (α) Evaluate the integral $I_n = \int e^{-x}x^n\,dx$. Now, $I_0 = \int e^{-x}\,dx = -e^{-x}$. Moreover, integrating by parts, we have (for $n > 0$):
$$I_n = -\int x^n d(e^{-x}) = -x^n e^{-x} + \int e^{-x}\frac{dx^n}{dx}\,dx$$
$$= -x^n e^{-x} + nI_{n-1}.$$

In particular, $I_1 = -xe^{-x}+I_0$, $I_2 = -x^2e^{-x}+2I_1$, ... Multiplying I_0 by $n!$, I_1 by $\frac{n!}{1!}$, ..., I_{n-1} by $\frac{n!}{(n-1)!}$, successively, and adding, we obtain

$$\int e^{-x}x^n dx = -n! e^{-x}\left(1+\frac{x}{1!}+\frac{x^2}{2!}+\ldots+\frac{x^n}{n!}\right)+C.$$

(β) Let $I_n = \int \frac{dx}{(1+x^2)^n}$. Now, $I_1 = \arctan x$. Moreover, we have for $n > 1$

$$I_n = \int \frac{1+x^2-x^2}{(1+x^2)^n}dx = I_{n-1} - \int \frac{x^2 dx}{(1+x^2)^n}.$$

Since

$$\frac{d}{dx}\frac{1}{(1+x^2)^{n-1}} = -2(n-1)\frac{x}{(1+x^2)^n},$$

we obtain integrating by parts

$$\int \frac{x^2 dx}{(1+x^2)^n} = \frac{-1}{2n-2}\int xd\frac{1}{(1+x^2)^{n-1}}$$

$$= \frac{-1}{2n-2}\left[\frac{x}{(1+x^2)^{n-1}} - I_{n-1}\right].$$

Consequently,

$$I_n = I_{n-1} + \frac{1}{2n-2}\cdot\frac{x}{(1+x^2)^{n-1}} - \frac{1}{2n-2}I_{n-1}$$

$$= \frac{1}{2n-2}\cdot\frac{x}{(1+x^2)^{n-1}} + \frac{2n-3}{2n-2}I_{n-1}.$$

This is the required recurrence formula.

(γ) Let $y = f(x)$, $z = g(x)$. We calculate the integral $\int zy^{(n+1)}dx$ applying n times the integration by parts:

$$\int zy^{(n+1)}dx = zy^{(n)} - \int z'y^{(n)}dx,$$
$$\int z'y^{(n)}dx = z'y^{(n-1)} - \int z''y^{(n-1)}dx,$$
$$\ldots\ldots\ldots\ldots\ldots\ldots\ldots\ldots$$
$$\int z^{(n)}y' dx = z^{(n)}y - \int z^{(n+1)}y\, dx.$$

Hence
$$\int zy^{(n+1)} dx =$$
$$= zy^{(n)} - z'y^{(n-1)} + \ldots + (-1)^n z^{(n)} y + (-1)^{n+1} \int z^{(n+1)} y \, dx \, .$$

This formula may be applied e.g. to the previously calculated integral $\int e^{-x} x^n \, dx$ after taking into account that

$$e^{-x} = (-1)^k \frac{d^k(e^{-x})}{dx^k} \, .$$

9.4. Integration of rational functions

Let a rational function

$$f(x) = \frac{P(x)}{Q(x)}$$

be given, where $P(x)$ and $Q(x)$ are two polynomials. The integral $\int f(x) \, dx$ has to be evaluated.

We may assume that *the degree of the numerator is less than the degree of the denominator*. For, if this were not so, we could divide the numerator by the denominator and then we have to integrate the sum of a polynomial (eventually reduced to a constant) and of a fraction, the numerator and the denominator of which are polynomials, and the degree of the numerator is less than the degree of the denominator. The first of these two terms is easy to integrate (as a polynomial); thus our task is to integrate the second component, i.e. a quotient of the form $\frac{P(x)}{Q(x)}$, where the polynomial $P(x)$ is of a degree less than the degree of the polynomial $Q(x)$.

The following theorem on rational functions is proved in books on algebra:

Let us call an expression of the form

$$\frac{A}{(x-p)^k} \quad \text{or} \quad \frac{Cx+D}{[(x-q)^2+r^2]^k} \, ,$$

where A and p, or C, D, q and r are real numbers, a partial fraction.

Then a rational function $\frac{P(x)}{Q(x)}$ is a sum of partial fractions the denominators of which are the factors of the polynomial $Q(x)$.

To decompose a rational function into partial fractions, first the polynomial $Q(x)$ has to be factorized into prime factors. Then the coefficients appearing in numerators of the partial fractions have to be found. The factorization of the polynomial $Q(x)$ in factors is obtained applying the theorem known from algebra stating that if p is a (real or complex) root of the equation $Q(x) = 0$, then $x-p$ is a factor of the polynomial $Q(x)$; moreover, if p is a root of the multiplicity n, then $(x-p)^k$ is a factor of this polynomial for each $k \leqslant n$. Finally, if $u+iv$ is a complex root of the equation $Q(x) = 0$ with real coefficients, then $u-iv$ is a root of this equation, too (of the same multiplicity), and so the polynomial $Q(x)$ has the product $(x-u-iv)(x-u+iv) = (x-u)^2 + v^2$ as a factor.

The coefficients appearing in the numerators of partial fractions are usually found by the so called *method of the undetermined coefficients*, which we shall learn by examples.

The denominator of the rational function $\frac{x-1}{(x-2)^2(x-3)}$ possesses the factors $(x-2)^2$, $(x-2)$ and $(x-3)$. Thus, our rational function is of the form

$$\frac{x-1}{(x-2)^2(x-3)} = \frac{A}{(x-2)^2} + \frac{B}{x-2} + \frac{C}{x-3}.$$

To calculate the coefficients A, B and C, we reduce the fractions on the right side of the above equation to a common denominator. We obtain

$$\frac{x-1}{(x-2)^2(x-3)} =$$
$$= \frac{(B+C)x^2 + (A-5B-4C)x - 3A + 6B + 4C}{(x-2)^2(x-3)}.$$

The numerators on the left side and on the right side of the above identity are equal one to another for any value of x. Hence it follows that

$$B+C = 0, \quad A-5B-4C = 1, \quad -3A+6B+4C = -1.$$

These equations give the values A, B and C:

$$A = -1, \quad B = -2, \quad C = 2.$$

Consequently,

$$\frac{x-1}{(x-2)^2(x-3)} = \frac{-1}{(x-2)^2} - \frac{2}{x-2} + \frac{2}{x-3}.$$

We note that the coefficients A, B, and C might be found in a slightly simpler way as follows. After reducing to the common denominator we obtain the following identity:

$$A(x-3) + B(x-2)(x-3) + C(x-2)^2 = x-1.$$

Substituting in this identity $x = 2$, we obtain $A = -1$; the substitution $x = 3$ gives $C = 2$. Hence, substituting $x = 1$, we conclude $B = -2$.

Another example of the decomposition of a rational function in partial fractions is given by:

$$\frac{x-1}{(x^2+1)^2(x+1)} = \frac{Ax+B}{(x^2+1)^2} + \frac{Cx+D}{x^2+1} + \frac{E}{x+1}.$$

Reducing to the common denominator we obtain the identity

$$(Ax+B)(x+1) + (Cx+D)(x^2+1)(x+1) + E(x^2+1)^2 = x-1,$$

whence we easily find:

$$A = 1, \quad B = 0, \quad C = \tfrac{1}{2}, \quad D = -\tfrac{1}{2}, \quad E = -\tfrac{1}{2}.$$

The theorem on the decomposition of a rational function into partial fractions reduces the problem of integration of rational functions to the integration of partial fractions. Now, we shall see that *the partial fractions may be integrated by the methods which we have learnt already.*

Indeed, substituting $y = x-p$ we obtain

$$\int \frac{A\,dx}{(x-p)^k} = A \int \frac{dy}{y^k} = A \log y \quad \text{for} \quad k=1,$$

$$\text{or} \quad = \frac{A}{1-k} \cdot \frac{1}{y^{k-1}} \quad \text{for} \quad k>1, \quad \text{respectively.}$$

The evaluation of the integral

$$\int \frac{(Cx+D)\,dx}{[(x-q)^2+r^2]^k}$$

is reduced to the evaluation of two integrals:

$$\int \frac{dx}{[(x-q)^2+r^2]^k} \quad \text{and} \quad \int \frac{(x-q)\,dx}{[(x-q)^2+r^2]^k}.$$

Substituting $y = x-q$, the first of these integrals is reduced to the integral $\int \frac{dy}{(y^2+r^2)^k}$, which is reduced by the substitution $z = y/r$ to the integral $\int \frac{dz}{(z^2+1)^k}$ which we already know (cf. § 9.3, 5 (β)).

The second integral we find by the formula § 9.3' (ζ) and (η) substituting $y = (x-q)^2+r^2$; namely,

$$\int \frac{(x-q)\,dx}{[(x-q)^2+r^2]^k} = \frac{1}{2} \log[(x-q)^2+r^2]$$

$$\text{or} \quad = \frac{1}{2(k-1)} \cdot \frac{1}{[(x-q)^2+r^2]^{k-1}}$$

according to whether $k=1$ or $k \neq 1$.

Hence, the integration of rational functions can always be carried out (of course, if the roots of the equation $Q(x) = 0$ are known).

EXAMPLES. (α) $I = \int \frac{dx}{x^2+x+1}$. Since

$$x^2+x+1 = x^2+x+\tfrac{1}{4}+\tfrac{3}{4} = (x+\tfrac{1}{2})^2+\tfrac{3}{4},$$

so substituting first $y = x + \frac{1}{2}$ and then $z = \frac{2}{\sqrt{3}} y$, we obtain

$$I = \int \frac{dy}{y^2 + \frac{3}{4}} = \int \frac{\frac{4}{3} dy}{(\frac{4}{3} y^2 + 1)} = \frac{2}{\sqrt{3}} \int \frac{dz}{z^2 + 1} = \frac{2}{\sqrt{3}} \arctan z$$
$$= \frac{2}{\sqrt{3}} \arctan \frac{2}{\sqrt{3}} \left(x + \frac{1}{2} \right).$$

(β) $\quad I = \int \frac{3x^3 + 10x^2 - x}{(x^2 - 1)^2} dx$

$$= \int \left[\frac{A}{(x-1)^2} + \frac{B}{x-1} + \frac{C}{(x+1)^2} + \frac{D}{x+1} \right] dx.$$

Reducing the right side of this equation to the common denominator, we easily find: $A = 3$, $B = 4$, $C = 2$, $D = -1$. Integrating, we obtain

$$I = -\frac{3}{x-1} + 4 \log |x-1| - \frac{2}{x+1} - \log |x+1|$$
$$= \frac{5x + 1}{1 - x^2} + \log \frac{(x-1)^4}{|x+1|}.$$

Remark. Analysing the method of integration of rational functions, we state that the integral of a rational function is of the form

$$W(x) + A \log U(x) + B \arctan V(x),$$

where $W(x)$, $U(x)$ and $V(x)$ are rational functions and A and B are constant coefficients.

9.5. Integration of irrational functions of the second degree

We shall consider an integral of the form

$$I = \int \sqrt{ax^2 + bx + c} \, dx.$$

First of all, let us consider the case when $a = 0$. Then we may assume that $b \neq 0$, since otherwise the function under the sign of the integral would be reduced to a constant. Hence, we have to evaluate the integral

$$I = \int \sqrt{bx + c} \, dx, \quad b \neq 0.$$

We substitute $t = bx+c$, i.e. $x = \dfrac{t-c}{b}$, whence $\dfrac{dx}{dt} = \dfrac{1}{b}$. Then the integral is transformed into the following integral:

$$I = \int \sqrt{t}\frac{dx}{dt}dt = \int \sqrt{t}\frac{1}{b}dt = \frac{2}{3b}t\sqrt{t} = \frac{2}{3b}(bx+c)\sqrt{bx+c}.$$

Now, let us assume that $a \neq 0$. We consider two cases:

(1) $b^2 - 4ac \geqslant 0$. As is known, this assumption implies that the equation $ax^2+bx+c = 0$ possesses real roots and the trinomial of second degree is of the form $ax^2+bx+c = a(x-p)(x-q)$, where p and q are the roots of the above equation. Thus, we have

$$I = \int \sqrt{ax^2+bx+c}\,dx = \int (x-p)\sqrt{a\frac{x-q}{x-p}}\,dx.$$

Let us substitute $a\dfrac{x-q}{x-p} = t^2$, i.e. $x = \dfrac{aq-t^2p}{a-t^2}$.
We obtain

$$I = \int t\frac{aq-ap}{a-t^2}\cdot\frac{dx}{dt}dt.$$

This integral being an integral of a rational function (since $\dfrac{dx}{dt}$ is a rational function, as is easily seen), it may be calculated by the methods given in § 9.4.

(2) $b^2-4ac < 0$. In this case it may be assumed that $a > 0$, since otherwise the trinomial constituting the function under the sign of the root, which is of the form $a[(x-\alpha)^2+\beta^2]$, would be always negative. We apply the following Euler substitution:

$$\sqrt{ax^2+bx+c} = x\sqrt{a}+t, \quad \text{i.e.} \quad bx+c = 2xt\sqrt{a}+t^2,$$

that is

$$x = \frac{t^2-c}{b-2t\sqrt{a}}.$$

Since $\sqrt{ax^2+bx+c}$ as well as $\dfrac{dx}{dt}$ are rational functions of the variable t, so our integral is reduced to an integral of a rational function.

The above Euler substitution may be applied also in the case (1), if only $a > 0$.

If $c > 0$, the second Euler substitution may be applied:
$$\sqrt{ax^2+bx+c} = xt + \sqrt{c}, \quad \text{i.e.} \quad ax+b = xt^2 + 2t\sqrt{c},$$
that is
$$x = \frac{2t\sqrt{c}-b}{a-t^2}.$$

Similar to the previous case, we see that this substitution reduces the integral I to an integral of a rational function with respect to the variable t.

Consequently, we conclude that the integral I may be evaluated for all values of the constants a, b, and c. It follows also from the above considerations easily that if $R(u, v)$ is a rational function of the variable u and v, then the integral
$$\int R\left(\sqrt{ax^2+bx+c},\, x\right) dx$$
is reduced to the integration of a rational function (thus, the integration is performable).

EXAMPLES. (α) Let $I = \dfrac{dx}{\sqrt{x^2+c}}$ for $x^2 > -c$. We shall apply the first Euler substitution in a slightly modified form. Namely, let us substitute $\sqrt{x^2+c} = t-x$, whence $x = \dfrac{t^2-c}{2t}$ and $\dfrac{dx}{dt} = \dfrac{t^2+c}{2t^2}$.

Hence $t-x = \dfrac{t^2+c}{2t}$ and
$$I = \int \frac{2t}{t^2+c} \cdot \frac{t^2+c}{2t^2} dt = \int \frac{dt}{t}$$
$$= \log|t| + C = \log\left|x + \sqrt{x^2+c}\right| + C.$$

(β) $I = \int \sqrt{x^2+c}\,dx$ for $x^2 > -c$. This integral may be evaluated by means of the Euler substitution. However, it may be calculated more simply by reduction (by integration by parts) to the previous integral. Namely,

$$I = x\sqrt{x^2+c} - \int x\frac{d\sqrt{x^2+c}}{dx}dx = x\sqrt{x^2+c} - \int \frac{x^2\,dx}{\sqrt{x^2+c}}$$

$$= x\sqrt{x^2+c} - \int \frac{x^2+c-c}{\sqrt{x^2+c}}dx = x\sqrt{x^2+c} - I + c\int \frac{dx}{\sqrt{x^2+c}}.$$

Hence,

$$I = \frac{1}{2}x\sqrt{x^2+c} + \frac{c}{2}\log\left|x+\sqrt{x^2+c}\right| + C.$$

Let us note that the formulae for the derivatives of the functions inverse to the hyperbolic functions (§ 7, (22) and (23)) make it possible to write directly the formulae for the integrals considered in the example (α) for $c = \pm 1$. Namely,

$$\int \frac{dx}{\sqrt{x^2+1}} = \operatorname{ar\,sinh} x + C = \log\left|x+\sqrt{x^2+1}\right| + C,$$

$$\int \frac{dx}{\sqrt{x^2-1}} = \operatorname{ar\,cosh} x + C = \log\left|x+\sqrt{x^2-1}\right| + C \text{ for } x > 1.$$

Remark. As we have seen, the integration of irrational functions of second degree is always performable. However, replacing the quadratic ax^2+bx+c under the sign of the root by a polynomial of the third or higher degree, the problem of integrability becomes in general (i.e. without special assumptions on the coefficients) impossible to solve in the domain of elementary functions.

In particular, the integral

$$\int \frac{dx}{\sqrt{(1-x^2)(1-k^2x^2)}}$$

cannot be expressed by means of elementary functions. Thus, it leads outside the domain of elementary functions (similarly as the integral $\int \frac{dx}{x}$ leads outside the domain of rational functions). It is the so-called *elliptic integral*.

Similarly, the integrals

$$\int \frac{dx}{\sqrt{\cos 2x}}, \quad \int \frac{dx}{\sqrt{1-k^2\sin^2 x}}$$

cannot be expressed by means of elementary functions (and lead also to elliptic integrals).

9.6. Integration of trigonometric functions

Let $R(u, v)$ denote a rational function of two variables u and v. The method of integration of the function $R(\sin x, \cos x)$ is based on the substitution $t = \tan \tfrac{1}{2} x$.

We conclude from the identity $\frac{1}{\cos^2 \tfrac{1}{2} x} = 1 + \tan^2 \tfrac{1}{2} x$, that $\cos^2 \tfrac{1}{2} x = \frac{1}{1+t^2}$ and, since $\cos x = 2\cos^2 \tfrac{1}{2} x - 1$,

$$\cos x = \frac{2}{1+t^2} - 1 = \frac{1-t^2}{1+t^2}.$$

At the same time,

$$\sin^2 \tfrac{1}{2} x = 1 - \cos^2 \tfrac{1}{2} x = \frac{t^2}{1+t^2},$$

whence

$$\sin x = 2\sin \tfrac{1}{2} x \cos \tfrac{1}{2} x = \frac{2t}{1+t^2}.$$

Finally, the equation $x = 2\arctan t$ implies

$$\frac{dx}{dt} = \frac{2}{1+t^2}.$$

The above three formulae give

$$\int R(\sin x, \cos x)\, dx = \int R\left(\frac{2t}{1+t^2}, \frac{1-t^2}{1+t^2}\right) \cdot \frac{2}{1+t^2}\, dt,$$

and the last integral, being an integral of a rational function of variable t, may be calculated by methods with which we have already become acquainted.

EXAMPLES. (α) Let $I = \int \dfrac{dx}{\cos x}$. Applying the substitution $t = \tan\tfrac{1}{2}x$ we obtain

$$I = \int \frac{1+t^2}{1-t^2} \cdot \frac{2}{1+t^2}\, dt = \int \frac{2dt}{1-t^2}$$
$$= \int \frac{dt}{1+t} + \int \frac{dt}{1-t} = \log\left|\frac{1+t}{1-t}\right|.$$

But $\dfrac{1+t}{1-t} = \dfrac{1+\tan\tfrac{1}{2}x}{1-\tan\tfrac{1}{2}x} = \tan(\tfrac{1}{2}x + \tfrac{1}{4}\pi)$. Consequently,

$$\int \frac{dx}{\cos x} = \log|\tan(\tfrac{1}{2}x + \tfrac{1}{4}\pi)|.$$

Hence

$$\int \frac{dx}{\sin x} = \log|\tan\tfrac{1}{2}x|,$$

since $\sin x = \cos(x - \tfrac{1}{2}\pi)$ and so

$$\int \frac{dx}{\sin x} = \int \frac{d(x - \tfrac{1}{2}\pi)}{\cos(x - \tfrac{1}{2}\pi)} = \log|\tan\tfrac{1}{2}x|.$$

The last integral may also be deduced easily from the integrals (cf. § 9.3, (ε)):

$$\int \tan x\, dx = -\log|\cos x|, \quad \int \cot x\, dx = \log|\sin x|,$$

since

$$\int \frac{dx}{\sin x} = \int \frac{dx}{2\sin\tfrac{1}{2}x \cos\tfrac{1}{2}x} = \frac{1}{2}\int \frac{\sin^2\tfrac{1}{2}x + \cos^2\tfrac{1}{2}x}{\sin\tfrac{1}{2}x \cos\tfrac{1}{2}x}\, dx$$
$$= \int \tan\tfrac{1}{2}x\, d(\tfrac{1}{2}x) + \int \cot\tfrac{1}{2}x\, d(\tfrac{1}{2}x)$$
$$= -\log|\cos\tfrac{1}{2}x| + \log|\sin\tfrac{1}{2}x| = \log|\tan\tfrac{1}{2}x|.$$

If we know the integral $\int \dfrac{dx}{\sin x}$, we find immediately the integral $\int \dfrac{dx}{\cos x}$, substituting $\cos x = \sin(x + \tfrac{1}{2}\pi)$.

(β) The irrational functions of second degree may be reduced easily to trigonometric integrals of the form

$R(\sin x, \cos x)$. Namely, the functions
$$\sqrt{1-x^2}, \quad \sqrt{x^2+1}, \quad \sqrt{x^2-1}$$
may be reduced to the functions of this type by substitutions $x = \sin t$, $x = \tan t$, $x = \sec t$; in any of these three cases the derivative $\dfrac{dx}{dt}$ is also a rational function of the functions $\sin t$ and $\cos t$. Namely,

$$\sqrt{1-\sin^2 t} = \cos t, \qquad \frac{d\sin t}{dt} = \cos t,$$

$$\sqrt{\tan^2 t + 1} = \sec t = \frac{1}{\cos t}, \qquad \frac{d\tan t}{dt} = \frac{1}{\cos^2 t},$$

$$\sqrt{\sec^2 t - 1} = \tan t = \frac{\sin t}{\cos t}, \qquad \frac{d\sec t}{dt} = \frac{\sin t}{\cos^2 t};$$

e.g.
$$\int \sqrt{x^2+1}\, dx = \int \frac{dt}{\cos^3 t}, \quad \text{where} \quad x = \tan t.$$

Sometimes it is more convenient to substitute $x = \cos t$ instead of $x = \sin t$ or $x = \cot t$ instead of $x = \tan t$; e.g. let us substitute $x = \cot t$ in the integral

$$\int \frac{dx}{\sqrt{x^2+1}} = \int \sin t \frac{d\cot t}{dt}\, dt = -\int \frac{dt}{\sin t} = -\log|\tan \tfrac{1}{2} t|.$$

Since $\dfrac{1}{x} = \tan t = \dfrac{2\tan \tfrac{1}{2} t}{1-\tan^2 \tfrac{1}{2} t}$, so $\tan \tfrac{1}{2} t = -x + \sqrt{x^2+1}$. Hence we find the integral

$$\int \frac{dx}{\sqrt{x^2+1}} = \log\left(x + \sqrt{x^2+1}\right),$$

calculated previously in another way.

Now, we shall give some other types of integrals of trigonometric functions, being of importance in applications.

The integral $I_n = \int \sin^n x\, dx$.

We shall give the recurrence formula.
$$I_n = -\int \sin^{n-1} x \, d\cos x = -\sin^{n-1} x \cos x + \int \cos x \frac{d\sin^{n-1} x}{dx} dx$$
$$= -\sin^{n-1} x \cos x + (n-1) \int \cos^2 x \sin^{n-2} x \, dx$$
$$= -\sin^{n-1} x \cos x + (n-1) I_{n-2} - (n-1) I_n,$$
since $\cos^2 x = 1 - \sin^2 x$.

Hence the required formula:
$$I_n = -\frac{1}{n} \sin^{n-1} x \cos x + \frac{n-1}{n} I_{n-2} \qquad (\text{for } n \geqslant 2).$$

Similarly, we find
$$\int \cos^n x \, dx = \frac{1}{n} \cos^{n-1} x \sin x + \frac{n-1}{n} \int \cos^{n-2} x \, dx \qquad (n \geqslant 2).$$

The integrals
$$\int \sin mx \sin nx \, dx, \qquad \int \sin mx \cos nx \, dx, \qquad \int \cos mx \cos nx \, dx$$
can be evaluated applying the formulae
$$\sin mx \sin nx = \tfrac{1}{2} [\cos(m-n)x - \cos(m+n)x],$$
$$\sin mx \cos nx = \tfrac{1}{2} [\sin(m+n)x + \sin(m-n)x],$$
$$\cos mx \cos nx = \tfrac{1}{2} [\cos(m+n)x + \cos(m-n)x],$$
known from the trigonometry.

We find easily:
$$\int \sin mx \sin nx \, dx =$$
$$= \begin{cases} \dfrac{1}{2} \left[\dfrac{\sin(m-n)x}{m-n} - \dfrac{\sin(m+n)x}{m+n} \right] & \text{for } m \neq \pm n, \\ \dfrac{1}{2} \left(x - \dfrac{\sin 2mx}{2m} \right) & \text{for } m = n, \end{cases}$$

$$\int \sin mx \cos nx \, dx =$$
$$= \begin{cases} -\dfrac{1}{2} \left[\dfrac{\cos(m+n)x}{m+n} + \dfrac{\cos(m-n)x}{m-n} \right] & \text{for } m \neq \pm n, \\ -\dfrac{1}{2} \cdot \dfrac{\cos 2mx}{2m} & \text{for } m = n, \end{cases}$$

$$\int \cos mx \cos nx\, dx =$$
$$= \begin{cases} \dfrac{1}{2}\left[\dfrac{\sin(m+n)x}{m+n} + \dfrac{\sin(m-n)x}{m-n}\right] & \text{for } m \neq \pm n, \\ \dfrac{1}{2}\left(\dfrac{\sin 2mx}{2m} + x\right) & \text{for } m = n. \end{cases}$$

The integral $\int \dfrac{dx}{a\sin x + b\cos x}$.

To calculate this integral, we write the denominator of the function under the sign of the integral in the form $c\sin(x+\Theta)$. The values of c and Θ we find from the identity:
$$c\sin x \cos \Theta + c\cos x \sin \Theta = a\sin x + b\cos x,$$
i.e.
$$a = c\cos\Theta, \quad b = c\sin\Theta,$$
that is
$$c = \sqrt{a^2+b^2}, \quad \Theta = \arctan\frac{b}{a}.$$

We obtain
$$\int \frac{dx}{a\sin x + b\cos x} = \frac{1}{c}\int \frac{d(x+\Theta)}{\sin(x+\Theta)} = \frac{1}{c}\log\left|\tan\frac{x+\Theta}{2}\right|.$$

Remark. In § 9.2, 5 we gave the recurrence formula for the integral $\int \dfrac{dx}{(1+x^2)^n}$. This integral may be calculated also applying the formula for the integral $\int \cos^m x\, dx$. Namely, substituting $x = \tan t$, we obtain
$$\int \frac{dx}{(1+x^2)^n} = \int \cos^{2n} t \frac{d\tan t}{dt}\, dt = \int \cos^{2n-2} t\, dt.$$

Exercises on § 9

Find the indefinite integrals of the following functions:

1. $e^{ax}\sin x$,
2. xe^x,
3. $x\sin x$,
4. $x^a \log x$,
5. $\arctan x$,
6. $x^a(\log x)^m$,

7. $\dfrac{1}{\sin^2 x}$,   8. $\dfrac{1}{1+x^2+x^4}$,

9. $\dfrac{x^2-x+4}{(x^2-1)(x+2)}$,   10. $\dfrac{1}{\sqrt[3]{x^2}+\sqrt{x}}$,

11. $\dfrac{1}{\sqrt{x+1}-\sqrt{x-1}}$,   12. $\dfrac{1}{a^2+x^2}$,

13. $\dfrac{1}{\sqrt{x^2+c}}$,   substituting $\sqrt{x^2+c} = x+t$

(cf. § 9.5, Example (α)),

14. $\dfrac{x^m}{\sqrt{1-x^2}}$ (m an integer),

15. $\dfrac{x^m}{\sqrt{x^2+c}}$ (m an integer),

16. $\dfrac{ax+b}{(x^2+c^2)^n \sqrt{px^2+qx+r}}$.

Remark. Prove that if R is a rational function of two variables, then $R\left(x, \sqrt{ax^2+bx+c}\right) = W(x) + \dfrac{U(x)}{\sqrt{ax^2+bx+c}}$, where U and W are rational functions. Hence deduce (by 14-16) a general method for integration of irrational functions of second degree.

17. $\cot^n x$,   18. $\sec^n x$,   19. $\sin^m x \cos^n x$,

20. $\cos x \cdot \cos 2x \cdot \cos 3x$,   21. $\dfrac{\arcsin x}{x^3}$.

§ 10. DEFINITE INTEGRALS

10.1. Definition and examples

Let a function $y = f(x)$ continuous in a closed interval $a \leqslant x \leqslant b$ be given. By the *definite integral* $f(x)dx$ *from* a *to* b we understand

(1) $$\int_a^b f(x)\,dx = F(b) - F(a),$$

where F is an arbitrary primitive function of the function f in the interval ab, i.e. $F'(x) = f(x)$ for $a < x < b$, $F'_+(a) = f(a)$ and $F'_-(b) = f(b)$.

The definite integral does not depend on the choice of·the function F; this means that if G is also a primitive function of the function f, then

$$G(b) - G(a) = F(b) - F(a).$$

This follows immediately from the fact that the difference between the functions G and F is constant in the interval $a \leqslant x \leqslant b$ (cf. § 7.5, 4, Remark).

The formula (1) will be assumed also in the case when $b < a$. Hence we have

(2) $$\int_b^a f(x)\,dx = -\int_a^b f(x)\,dx$$

and

(3) $$\int_a^a f(x)\,dx = 0.$$

From Theorem 2, of § 9, 2 it follows immediately that *every function continuous in an interval* $a \leqslant x \leqslant b$ *possesses the integral* $\int_a^b f(x)\,dx$, that is, as we say, it is *integrable* in this interval.

To indicate the difference $F(b) - F(a)$, we use also the symbol

$$[F(x)]_a^b = F(b) - F(a).$$

Remark. It follows from the definition that the definite integral depends on the function f and on the limits of integration. However, it is not a function of the variable x (x appears here only as a variable of integration). The value of the integral will not be changed, if we replace the variable x by another variable, e.g. t:

$$\int_a^b f(x)\,dx = \int_a^b f(t)\,dt.$$

EXAMPLES. (α) $\int_a^b c\,dx = c(b-a)$. Indeed, we have
$\int c\,dx = cx + C$, whence $\int_a^b c\,dx = [cx]_a^b = c(b-a)$.

(β) $\int_0^1 x^2\,dx = \frac{1}{3}$, because $\int x^2\,dx = \frac{1}{3}x^3 + C$, whence $\int_0^1 x^2\,dx = [\frac{1}{3}x^3]_0^1 = \frac{1}{3}$.

(γ) $\int_0^{\pi/2} \sin x\,dx = [-\cos x]_0^{\pi/2} = -\cos(\pi/2) + \cos 0 = 1$.

(δ) The following definite integrals (Fourier's, cf. § 11.7) are easily found on the basis of the formulae for the indefinite integrals (§ 9.6): if m and n indicate two different positive integers, then

$$\int_{-\pi}^{\pi} \sin mx \sin nx\,dx = 0 = \int_{-\pi}^{\pi} \cos mx \cos nx\,dx\,;$$

for arbitrary positive integers m and n:

$$\int_{-\pi}^{\pi} \sin mx \cos nx\,dx = 0\,;$$

for arbitrary positive integer m:

$$\int_{-\pi}^{\pi} \sin^2 mx\,dx = \pi = \int_{-\pi}^{\pi} \cos^2 mx\,dx\,.$$

(ε) For an arbitrary positive integer n the formulae

(4) $\int_0^{\pi/2} \sin^{2n} x\,dx = \frac{1\cdot 3\cdot\ldots\cdot(2n-1)}{2\cdot 4\cdot\ldots\cdot 2n}\cdot\frac{\pi}{2}$,

$\int_0^{\pi/2} \sin^{2n+1} x\,dx = \frac{2\cdot 4\cdot\ldots\cdot 2n}{3\cdot 5\cdot\ldots\cdot(2n+1)}$

hold.

These formulae follow easily from the recurrence formula (§ 9.6):

$\int \sin^n x\,dx = -\frac{1}{n}\sin^{n-1} x\cos x + \frac{n-1}{n}\int \sin^{n-2} x\,dx \quad (n \geqslant 2)$

and from the formulae:

$$\left[-\frac{1}{n}\sin^{n-1}x\cos x\right]_0^{\pi/2} = 0,$$

$$\int_0^{\pi/2} \sin^1 x\,dx = 1, \quad \int_0^{\pi/2} \sin^0 x\,dx = \tfrac{1}{2}\pi.$$

Let us note that if we replace sine by cosine, then the formulae (4) remain true.

10.2. Calculation formulae

The first three general formulae for the indefinite integrals (§ 9.3) lead immediately to the following three calculation formulae for the definite integrals (we assume that the functions under the signs of the integral are continuous in the closed interval of integration):

1. $\int_a^b [f(x) \pm g(x)]\,dx = \int_a^b f(x)\,dx \pm \int_a^b g(x)\,dx,$

2. $\int_a^b c \cdot f(x)\,dx = c \int_a^b f(x)\,dx,$

3. $\int_a^b f(x)g'(x)\,dx = [f(x)g(x)]_a^b - \int_a^b f'(x)g(x)\,dx.$

The formula for integration by substitution leads to the following formula:

4. $\int_a^b g[f(x)]f'(x)\,dx = \int_{f(a)}^{f(b)} g(y)\,dy.$

Namely, let $G(y)$ be a primitive function of the function $g(y)$. Then (by formula 4 of § 9.3), $G[f(x)]$ is the primitive function of the function $g[f(x)]f'(x)$. Thus, according to the definition of the definite integral,

$$\int_a^b g[f(x)]f'(x)\,dx = G[f(b)] - G[f(a)] = \int_{f(a)}^{f(b)} g(y)\,dy.$$

5. The theorem on the division of the interval of integration:

$$\int_a^c f(x)\,dx + \int_c^b f(x)\,dx = \int_a^b f(x)\,dx.$$

Namely, denoting by F a primitive function of the function f, we have

$$[F(c)-F(a)] + [F(b)-F(c)] = F(b)-F(a).$$

6. *If* $f(x) \geqslant 0$ *and* $a < b$, *then* $\int_a^b f(x)\,dx \geqslant 0$.

Indeed, denoting by F a primitive function of the function f, we have $F'(x) = f(x) \geqslant 0$; hence it follows that the function F is increasing (in the wider sense). So $F(a) \leqslant F(b)$, i.e. $\int_a^b f(x)\,dx = F(b)-F(a) \geqslant 0$.

Formula 6 implies the following slightly more general formula:

7. *If* $f(x) \leqslant g(x)$ *and* $a < b$, *then* $\int_a^b f(x)\,dx \leqslant \int_a^b g(x)\,dx$.

By assumption, $g(x)-f(x) \geqslant 0$. Thus, by 6, we have $\int_a^b [g(x)-f(x)]\,dx \geqslant 0$, whence by 1,

$$\int_a^b g(x)\,dx - \int_a^b f(x)\,dx \geqslant 0.$$

As is seen from formula 7, the relation $\leqslant$ remains unchanged by the integration (when $a < b$). We shall deduce from 7 the following two formulae:

8. $\left| \int_a^b f(x)\,dx \right| \leqslant \int_a^b |f(x)|\,dx$, *if* $a < b$.

Namely, it follows (according to 7) by integration of the double inequality $-|f(x)| \leqslant f(x) \leqslant |f(x)|$:

$$\int_a^b -|f(x)|\,dx \leqslant \int_a^b f(x)\,dx \leqslant \int_a^b |f(x)|\,dx,$$

i.e.

$$-\int_a^b |f(x)|\,dx \leqslant \int_a^b f(x)\,dx \leqslant \int_a^b |f(x)|\,dx.$$

Hence we get the required formula (since the inequality $-u \leqslant v \leqslant u$ implies the inequality $|v| \leqslant u$ (cf. § 1, (17)).

9. The mean-value formula.

$$\int_a^{a+h} f(x)\,dx = hf(a + \Theta h),$$

where Θ is a suitably chosen number satisfying the condition $0 < \Theta < 1$.

Let F denote a primitive function of the function f, i.e. $F'(x) = f(x)$. Then we have

$$\int_a^{a+h} f(x)\,dx = F(a+h) - F(a) = hF'(a+\Theta h) = hf(a+\Theta h)$$

according to the mean-value theorem of the differential calculus (§ 7.5, (β)).

10. Let a function f be continuous in the interval $a \leqslant x \leqslant b$. Let us write

(1) $$g(x) = \int_a^x f(t)\,dt \ \ (^1).$$

Then the following basic formula holds:

(2) $$g'(x) = f(x), \quad \text{i.e.} \quad \frac{d}{dx}\int_a^x f(t)\,dt = f(x),$$

(¹) It may be also written $g(x) = \int_a^x f(x)\,dx$.

i.e. the integral $\int_a^x f(t)dt$, considered as a function of the upper limit of integration, is a primitive function of the function under the sign of the integral.

To prove this, let F denote an arbitrary primitive function of the function f, i.e. let $F'(x) = f(x)$. According to the formula (1) of § 10.1, we have $g(x) = F(x) - F(a)$. Differentiating, we obtain $g'(x) = F'(x) = f(x)$.

Remarks. (α) At the ends of the interval of integration we have

$$g'_+(a) = f(a) \quad \text{and} \quad g'_-(b) = f(b).$$

(β) Let us note that the function g is continuous as a differentiable function.

11. *If a sequence of continuous functions $f_n(x)$, $a \leqslant x \leqslant b$, is uniformly convergent to a function $f(x)$ in the interval ab, then the integral of the limit is equal to the limit of the integrals:*

(3) $$\int_a^b f(x)dx = \lim_{n=\infty} \int_a^b f_n(x)dx,$$

i.e. $$\int_a^b [\lim_{n=\infty} f_n(x)]dx = \lim_{n=\infty} \int_a^b f_n(x)dx.$$

Since the limit of a uniformly convergent sequence of continuous functions is a continuous function (§ 6.1, Theorem 1), the function f is integrable. Applying the formula 1 to the functions f and f_n, we have

(4) $$\int_a^b f(x)dx - \int_a^b f_n(x)dx = \int_a^b [f(x) - f_n(x)]dx.$$

At the same time, because of the uniform convergence of the sequence of functions $f_1, f_2, \ldots$ to the function f, there exists for a given $\varepsilon > 0$ a number k such that $|f(x) - f_n(x)| < \varepsilon$ for $n > k$. Hence, by (4), 8 and 7, we have

$$\left|\int_a^b f(x)\,dx - \int_a^b f_n(x)\,dx\right| =$$
$$= \left|\int_a^b [f(x)-f_n(x)]\,dx\right| \leqslant \int_a^b |f(x)-f_n(x)|\,dx \leqslant \varepsilon|b-a|.$$

In this way we have proved that for $n > k$ there holds the inequality

$$\left|\int_a^b f(x)\,dx - \int_a^b f_n(x)\,dx\right| \leqslant \varepsilon|b-a|.$$

This means that equation (3) is satisfied.

As a corollary of Theorem 11 we have:

12. $\displaystyle\int_a^b \left[\sum_{n=0}^\infty f_n(x)\right]dx = \sum_{n=0}^\infty \int_a^b f_n(x)\,dx,$

assuming the functions f_n to be continuous and the series $\sum_{n=0}^\infty f_n(x)$ to be uniformly convergent in the interval $a \leqslant x \leqslant b$.

EXAMPLES AND APPLICATIONS. (α) To evaluate the integral $\int_0^1 (1-x^2)^n\,dx$, we substitute $x = \sin t$. Then we have $0 = \sin 0$, $1 = \sin(\pi/2)$. Thus (cf. § 10.1, 5),

(5) $\displaystyle\int_0^1 (1-x^2)^n\,dx = \int_0^{\pi/2} \cos^{2n} t \, \frac{dx}{dt}\,dt$
$$= \int_0^{\pi/2} \cos^{2n+1} t\,dt = \frac{2\cdot 4\cdot\ldots\cdot 2n}{3\cdot 5\cdot\ldots\cdot(2n+1)}.$$

(β) $\displaystyle\lim_{n=\infty} \int_a^b \cos nx\,dx = 0 = \lim_{n=\infty}\int_a^b \sin nx\,dx.$

Indeed, substituting $y = nx$, we obtain

$$\int_a^b \cos nx\,dx = \frac{1}{n}\int_{na}^{nb} \cos y\,dy = \frac{1}{n}(\sin nb - \sin na).$$

Since $|\sin nb - \sin na| \leqslant 2$, we have $\lim\limits_{n=\infty} \int_a^b \cos nx\,dx = 0$.

Analogously it is proved that $\lim\limits_{n=\infty} \int_a^b \sin nx\,dx = 0$.

The following more general formula holds:

(γ) $\lim\limits_{n=\infty} \int_a^b f(x)\cos nx\,dx = 0 = \lim\limits_{n=\infty} \int_a^b f(x)\sin nx\,dx$,

where f is an arbitrary continuous function.

Let a number $\varepsilon > 0$ be given. According to the uniform continuity of the function f in the interval $a \leqslant x \leqslant b$ there exists an m such that dividing the segment ab in m equal segments, to any pair of points x and x' belonging to the very same from among these m segments the inequality $|f(x) - f(x')| < \varepsilon$ holds. Let $a_1, a_2, \ldots, a_{m-1}$ be the points of the division; moreover, let $a_0 = a$, $a_m = b$. By formula 5, we have

$$\int_a^b f(x)\cos nx\,dx = \sum_{k=1}^m \int_{a_{k-1}}^{a_k} f(x)\cos nx\,dx$$

$$= \sum_{k=1}^m \left\{ f(a_k) \int_{a_{k-1}}^{a_k} \cos nx\,dx + \int_{a_{k-1}}^{a_k} [f(x) - f(a_k)]\cos nx\,dx \right\}.$$

Thus (cf. formula 8),

$$\left|\int_a^b f(x)\cos nx\,dx\right| \leqslant \sum_{k=1}^m |f(a_k)| \cdot \left|\int_{a_{k-1}}^{a_k} \cos nx\,dx\right| + \varepsilon(b-a),$$

because

$$|f(x) - f(a_k)||\cos nx| \leqslant |f(x) - f(a_k)| < \varepsilon.$$

Since, as we have proved,

$$\left|\int_{a_{k-1}}^{a_k} \cos nx\, dx\right| = \frac{1}{n}|\sin na_k - \sin na_{k-1}| \leqslant \frac{2}{n},$$

so denoting by M the upper bound of the function $|f(x)|$ in the interval $a \leqslant x \leqslant b$, we obtain

$$\left|\int_a^b f(x)\cos nx\, dx\right| \leqslant \frac{2}{n} Mm + \varepsilon(b-a).$$

Whence choosing n so large that $\frac{2}{n} Mm < \varepsilon$ we have

$$\left|\int_a^b f(x)\cos nx\, dx\right| < \varepsilon(1+b-a).$$

Hence the first of the equations (γ) follows. The proof of the second one is analogous.

(δ) If the indefinite integral of a given function is not known but this function may be expanded in a uniformly convergent series of functions which we are able to integrate (e. g. in a power series), then the application of Theorem 12 makes it possible to express the definite integral by a series.

In this way we shall calculate the elliptic integral (cf. § 9.5, Remark):

$$I = \int_0^{\pi/2} \frac{dx}{\sqrt{1-k^2\sin^2 x}} \quad (k^2 < 1).$$

As we know (cf. § 8.4):

$$\frac{1}{\sqrt{1-t^2}} = 1 + \frac{1}{2}t^2 + \frac{1\cdot 3}{2\cdot 4}t^4 + \frac{1\cdot 3\cdot 5}{2\cdot 4\cdot 6}t^6 + \ldots \quad (|t|<1).$$

Hence, replacing t by $k\sin x$ we have:

$$\frac{1}{\sqrt{1-k^2\sin^2 x}} = 1 + \frac{1}{2}k^2\sin^2 x + \frac{1\cdot 3}{2\cdot 4}k^4\sin^4 x +$$
$$+ \frac{1\cdot 3\cdot 5}{2\cdot 4\cdot 6}k^6\sin^6 x + \ldots$$

This series is uniformly convergent in the interval $0 \leqslant x \leqslant \pi/2$ (even on the whole X-axis), since the series representing $\dfrac{1}{\sqrt{1-t^2}}$ is uniformly convergent in the interval $-|k| \leqslant t \leqslant |k|$. Integrating term by term and applying the formula ((4) of § 10.1)

$$\int_0^{\pi/2} \sin^{2n} x \, dx = \frac{1 \cdot 3 \cdot \ldots \cdot (2n-1)}{2 \cdot 4 \cdot \ldots \cdot 2n} \cdot \frac{\pi}{2},$$

we obtain

$$\int_0^{\pi/2} \frac{dx}{\sqrt{1-k^2 \sin^2 x}} =$$

$$= \frac{\pi}{2}\left[1 + \left(\frac{1}{2}\right) k^2 + \left(\frac{1 \cdot 3}{2 \cdot 4}\right)^2 k^4 + \left(\frac{1 \cdot 3 \cdot 5}{2 \cdot 4 \cdot 6}\right)^2 k^6 + \ldots \right].$$

10.3. Definite integral as a limit of sums

Now, we shall prove theorems which give important applications of the definite integral.

1. *Let a continuous function $y = f(x)$, $a \leqslant x \leqslant b$, and a number $\varepsilon > 0$ be given. Then there exists a number $\delta > 0$ possessing the following property: if the points $a_0 < a_1 < \ldots < a_m$, where $a_0 = a$ and $a_m = b$, satisfy the inequality $a_k - a_{k-1} < \delta$ for each $k = 1, 2, \ldots, m$ and if the points $x_1, \ldots, x_m$ are chosen from the intervals $a_0 a_1, \ldots, a_{m-1} a_m$, respectively (i.e. $a_{k-1} \leqslant x_k \leqslant a_k$), then*

$$\left| \sum_{k=1}^{m} f(x_k)(a_k - a_{k-1}) - \int_a^b f(x) \, dx \right| < \varepsilon.$$

Indeed, according to the uniform continuity of the function f in the interval $a \leqslant x \leqslant b$ (cf. § 6.4, 1) there exists a $\delta > 0$ such that the condition $|x - x'| < \delta$ implies

$|f(x)-f(x')| < \dfrac{\varepsilon}{b-a}$. By the formulae 5 and 9, § 10.2, we have

$$\int_a^b f(x)\,dx = \int_{a_0}^{a_1} f(x)\,dx + \ldots + \int_{a_{m-1}}^{a_m} f(x)\,dx$$
$$= f(x_1')(a_1-a_0) + \ldots + f(x_m')(a_m-a_{m-1})$$
$$= \sum_{k=1}^{m} f(x_k')(a_k-a_{k-1}),$$

where x_k' is a suitably chosen number in the interval $a_{k-1}a_k$.

Since the points x_k and x_k' belong to the interval $a_{k-1}a_k$, the distance between these points is not greater than the length of this interval, i.e. $|x_k-x_k'| \leqslant a_k-a_{k-1} < \delta$. Hence it follows that $|f(x_k)-f(x_k')| < \dfrac{\varepsilon}{b-a}$. Consequently

$$\left| \sum_{k=1}^{m} f(x_k)(a_k-a_{k-1}) - \int_a^b f(x)\,dx \right| =$$
$$= \left| \sum_{k=1}^{m} [f(x_k)-f(x_k')](a_k-a_{k-1}) \right|$$
$$\leqslant \sum_{k=1}^{m} |f(x_k)-f(x_k')|(a_k-a_{k-1})$$
$$< \dfrac{\varepsilon}{b-a} \sum_{k=1}^{m} (a_k-a_{k-1}) = \dfrac{\varepsilon}{b-a}(b-a) = \varepsilon.$$

In this way Theorem 1 is proved.

Now, let us consider instead of one partition of the segment ab (defined by the points $a_0, a_1, \ldots, a_m$), a sequence of partitions of this segment in smaller segments. Let the n-th partition be defined by the points

$a_{n,0}, a_{n,1}, \ldots, a_{n,l_n}$, where $a_{n,0} = a$, $a_{n,l_n} = b$.

Let r_n denote the length of the greatest segment of the n-th partition. We say that the above sequence of partitions is *normal*, if $\lim\limits_{n=\infty} r_n = 0$. The following theorem holds:

2. *Let a continuous function* $y = f(x)$, $a \leqslant x \leqslant b$, *and a normal sequence of partitions of this segment be given. If the point* $x_{n,k}$ *belongs to the k-th interval of the n-th partition* (*i.e.* $a_{n,k-1} \leqslant x_{n,k} \leqslant a_{n,k}$ *for* $k = 1, 2, \ldots, l_n$; $n = 1, 2, \ldots$), *then*

$$(6) \quad \int_a^b f(x)\,dx = \lim_{n=\infty} \sum_{k=1}^{l_n} f(x_{n,k})(a_{n,k} - a_{n,k-1}).$$

Let an $\varepsilon > 0$ be given. We choose N in such a way that for $n > N$ the inequality $r_n < \delta$ holds, i. e. that all intervals of the n-th partition have the length $< \delta$ (where δ has the same meaning as in the previous theorem). Then by Theorem 1 we have

$$\left| \sum_{k=1}^{l_n} f(x_{n,k})(a_{n,k} - a_{n,k-1}) - \int_a^b f(x)\,dx \right| < \varepsilon.$$

This means that equality (6) is satisfied.

It follows in particular that if all segments of the n-th partition are of the same length, namely, $2^{-n}(b-a)$, then

$$\int_a^b f(x)\,dx = (b-a)\lim_{n=1} 2^{-n} \sum_{k=1}^{2^n} f(a_{n,k}).$$

Remark. The notation $\int_a^b y\,dx$ is closely related to Theorem 2, so from the point of view of notions as from the historical point of view. The symbol $\int$, the modified letter S, had to indicate the summation of infinitely many "infinitesimal" terms of the form $y\,dx$; this concept, which was not well-defined from the logical point of view, becomes now expressed strictly in Theorem 2. It is easily seen that this theorem represents the definite integral not as an infinite sum but as a limit of sums of an increasing number of terms.

In § 10.11 we shall turn to the definition of the integral based on formula (6).

10.4. The integral as an area

Let $f(x) \geqslant 0$ for $a \leqslant x \leqslant b$.

The integral $\int_a^b f(x)dx$ is the area of the region P consisting of the points x, y of the plane satisfying the conditions:

$$a \leqslant x \leqslant b, \quad 0 \leqslant y \leqslant f(x).$$

It is really so, when the function f is linear. Then the region P is a trapezoid the bases of which (parallel to the Y-axis) have the length $f(a)$ and $f(b)$, respectively, and the hight has the length $b-a$. Thus the area of the trapezoid is $\frac{f(a)+f(b)}{2}(b-a)$.

On the other hand, the function f as a linear function is of the form $f(x) = cx+d$. Hence

$$\int_a^b f(x)dx = \tfrac{1}{2}c(b^2-a^2) + d(b-a) = (b-a)(\tfrac{1}{2}cb + \tfrac{1}{2}ca + d)$$

$$= \frac{f(a)+f(b)}{2}(b-a),$$

since $f(a) = ca+d$ and $f(b) = cb+d$.

Thus we see that in the case when the function f is linear, the notion of the area known from the elementary geometry is equivalent to the above defined notion of the area as the definite integral. This equivalence holds also in a wider sense, if f is a polygonal function (cf. § 6.4), i.e. when the graph of the function is a polygonal line. Namely, the region between this polygonal line and the X-axis consists of a certain number of trapezoids and hence its area is the sum of the areas of these trapezoids. Yet since, as we have stated above, the area of each of these trapezoids is a suitable integral, so the area of the whole region is the sum of these integrals, i.e. $\int_a^b f(x)dx$ (according to § 10.2, 5).

In the general case if f is an arbitrary continuous function, the function f is the limit of a uniformly convergent sequence of polygonal functions (cf. § 6.4): $f(x) = \lim_{n=\infty} f_n(x)$. Hence, by § 10.2, 11,

$$\int_a^b f(x)\,dx = \lim_{n=\infty} \int_a^b f_n(x)\,dx.$$

Then we see that the area of the region determined by the curve $y = f(x)$ (defined as the integral $\int_a^b f(x)\,dx$) is the limit of areas of polygons "inscribed" in this curve (cf. Fig. 10 in § 6.4).

Thus, there is here a complete analogy to the area of a circle, defined as the limit of areas of inscribed (or circumscribed) polygons. Similarly, we *define* the area of the region P given by the curve $y = f(x)$ as the integral $\int_a^b f(x)\,dx$.

Remark. This area could also be defined on the basis of the measure theory, which we shall not consider here. Then this definition could be proved to be equivalent to the definition of the area by means of the definite integral (we have performed such a proof in the most elementary case, namely, if f is linear or polygonal).

EXAMPLES. (α) Let $y = \sqrt{r^2 - x^2}$, $-r \leqslant x \leqslant r$. We shall calculate $\int_{-r}^{r} \sqrt{r^2 - x^2}\,dx$.

For this purpose we shall find the indefinite integral. Substituting $z = \dfrac{x}{r}$ we obtain (cf. § 9.3, 4)

$$\int \sqrt{r^2 - x^2}\,dx = r^2 \int \sqrt{1-z^2}\,dz = r^2\left(\frac{1}{2}\arcsin z + \frac{1}{2}z\sqrt{1-z^2}\right)$$

$$= r^2\left(\frac{1}{2}\arcsin\frac{x}{r} + \frac{1}{2}\cdot\frac{x}{r}\sqrt{1-\frac{x^2}{r^2}}\right).$$

Hence

$$\int_{-r}^{r} \sqrt{r^2 - x^2}\,dx = r^2\left(\tfrac{1}{2}\arcsin 1 - \tfrac{1}{2}\arcsin(-1)\right) = \tfrac{1}{2}\pi r^2,$$

since $\arcsin 1 = \tfrac{1}{2}\pi$ and $\arcsin(-1) = -\tfrac{1}{2}\pi$.

In this way we have obtained the formula for the area of the semicircle with a radius r, known from elementary geometry.

(β) The area of the region contained between the arc of the parabola $y = x^2$, the X-axis and the straight line $x = a$ is given by the integral $\int_0^a x^2 dx = \frac{1}{3}a^3$.

(γ) The area of the ellipse $\dfrac{x^2}{a^2} + \dfrac{y^2}{b^2} = 1$ will be calculated as follows: the above equation gives $y = b\sqrt{1 - \dfrac{x^2}{a^2}}$, whence the area of the ellipse is equal to $2b \int_{-a}^{a} \sqrt{1 - \dfrac{x^2}{a^2}}\, dx$. Applying, similarly as in the example (α), the formula for the indefinite integral $\int \sqrt{1-z^2}\, dz$ we find

$$2b \int_{-a}^{a} \sqrt{1 - \frac{x^2}{a^2}}\, dx = \pi ab.$$

(δ) The area between the sinusoid and the segment 0π on the X-axis is

$$\int_0^{\pi} \sin x\, dx = [-\cos x]_0^{\pi} = 2.$$

(ε) Let two continuous functions $u(x)$ and $v(x)$ defined in the segment $a \leqslant x \leqslant b$ be given. Moreover, let $u(x) < v(x)$ for $a < x < b$. The region N contained between these two curves, i. e. the set of points x, y on the plane satisfying the conditions:

$$a \leqslant x \leqslant b, \quad u(x) \leqslant y \leqslant v(x)$$

is called a *normal region*.

The area of the normal region N is $\int_a^b [v(x) - u(x)]\, dx$.

Indeed, if $u(x) \geqslant 0$ everywhere, then the region N is obtained from the region contained between the curve $y = v(x)$ and the X-axis by subtraction of the region contained between the curve $y = u(x)$ and the X-axis; the area of the first region is $\int_a^b v(x)dx$ and the area of the second one is $\int_a^b u(x)dx$. Hence the area N is the difference of these integrals.

Now, if the inequality $u(x) \geqslant 0$ does not hold for all x, then we denote by m the lower bound of the function u in the interval $a \leqslant x \leqslant b$ and by N_0 the normal region defined by the curves $u(x)-m$ and $v(x)-m$. The region N_0 is obtained from the region N by translation and so N_0 is congruent to N. Thus,

$$\text{area } N = \text{area } N_0 = \int_a^b \{[v(x)-m]-[u(x)-m]\}dx$$
$$= \int_a^b [v(x) - u(x)]dx.$$

Remarks. (α) Considering the area of the normal region it is meaningless whether we take its perimeter into account. Namely, we assume the perimeter to have the area 0. This follows from the fact that the curve $y = f(x)$ may be included in a normal region with an arbitrarily small area, namely, in the region contained between the arcs $y = f(x)+\varepsilon$ and $y = f(x)-\varepsilon$; the area of this region is $2\varepsilon(b-a)$.

(β) The symbol $y\,dx$ is called the *element of the area* (cf. § 7.13).

(γ) The integral $\int_a^b f(x)dx$ may be interpreted as an area also without the assumption $f(x) \geqslant 0$, if we agree upon considering the area of regions lying below the X-axis to be negative (where $a < b$). Thus, if the interval ab may be divided in smaller intervals in such a way that in everyone of these intervals the function is either everywhere nonnegative or nonpositive, then $\int_a^b f(x)dx$ is the algebraic sum of the areas defined by the arcs of the curve $y = f(x)$, in each of these intervals separately.

For example for the curve $y = \sin x$, $0 \leqslant x \leqslant 2\pi$, we divide the interval $0, 2\pi$ into two halves. The sine curve together with the interval $0, \pi$ of the X-axis gives a region with an area 2 and together with the interval $\pi, 2\pi$, a region with an area -2. The algebraic sum of these areas equals 0, i.e.

$$\int_0^{2\pi} \sin x \, dx = 0 .$$

(δ) Interpreting the definite integral as an area, many of the previously proved theorems become of a clear geometrical content (especially, if the function under the sign of the integral is nonnegative); e.g. Theorem 9 of § 10.2 means that the region defined by the arc $y = f(x)$, $a \leqslant x \leqslant b$, the X-axis and the perpendicular segments lying on the straight lines $x = a$ and $x = b$ has the same area as a rectangle having the segment $a \leqslant x \leqslant b$ as the base and the ordinate of a suitable chosen point on the arc $y = f(x)$, $a < x < b$, as the height.

It follows from Theorem 10, of § 10.2 (cf. remark (β)) that the area of a region defined by the arc $y = f(x)$, $a \leqslant x \leqslant x'$, varies continuously together with x'.

Theorem 11 of § 10.2 means that if the functions f_n are uniformly convergent in an interval ab to a function f, then the areas of the regions defined by f_n are convergent to the area of the region defined by f.

10.5. The length of an arc

Let a function $y = f(x)$, $a \leqslant x \leqslant b$, having a continuous derivative $f'(x)$ be given. The integral

(7) $$\int_a^b \sqrt{1 + \left(\frac{dy}{dx}\right)^2} \, dx$$

gives the length of the arc formed by the points x, y satisfying the conditions

(8) $\quad\quad\quad y = f(x), \quad a \leqslant x \leqslant b.$

It is indeed so, if f is a linear function, i.e. if the considered curve is a segment joining the points $[a, f(a)]$ and $[b, f(b)]$. The length of the segment is equal to the distance of its ends, i.e.

$$\sqrt{(b-a)^2 + [f(b)-f(a)]^2}.$$

Applying the integral we obtain the same result. Namely, let us write $f(x) = cx + d$. Hence $f(c) = c$. Thus,

$$\int_a^b \sqrt{1+c^2}\, dx = \sqrt{1+c^2}\,(b-a) = \sqrt{(b-a)^2 + [f(b)-f(a)]^2},$$

since

$$c = \frac{f(b)-f(a)}{b-a}.$$

In the general case (if we do not assume the linearity of the function f) we find the explanation of the above assumed definition of the length of an arc in the following construction.

We know (cf. § 6.4) that an arc of a curve given by the conditions (8) may be approximated by polygonal lines inscribed in this arc (cf. § 6.4, Fig. 10). Now, we shall prove that, denoting by L the length of this arc expressed by the integral (7) and by $L_1, L_2, \ldots$ the length of the successive polygonal lines, we have

$$L = \lim_{n=\infty} L_n.$$

Namely, let the n-th polygonal line be obtained by joining the points $[a_{n,0}, f(a_{n,0})], [a_{n,1}, f(a_{n,1})], \ldots, [a_{n,l_n}, f(a_{n,l_n})]$ by segments, successively, where $a_{n,0} = a$, $a_{n,l_n} = b$ and the length of the greatest interval of the n-th partition tends to 0 as n tends to ∞ (i.e. the sequence of these partitions is normal in the sense of § 10.3). The length

10. DEFINITE INTEGRALS

of the n-th polygonal line is the sum of the length of the particular segments, which constitute this polygonal line, i.e.

$$L_n = \sum_{n=1}^{l_n} \sqrt{(a_{n,k}-a_{n,k-1})^2+[f(a_{n,k})-f(a_{n,k-1})]^2}.$$

Applying the mean-value theorem we obtain

$$L_n = \sum_{k=1}^{l_n} (a_{n,k}-a_{n,k-1})\sqrt{1+[f'(x_{n,k})]^2},$$

where $a_{n,k-1} \leqslant x_{n,k} \leqslant a_{n,k}$.

Hence by Theorem 2 of § 10.3 (formula (6)), we conclude after substituting $\sqrt{1+[f'(x)]^2}$ in place of the function $f(x)$ that

$$\int_a^b \sqrt{1+[f'(x)]^2}\,dx = \lim_{n=\infty} L_n = L.$$

The remarks noted in the case of the definition of the area are valid in the case of the definition of the length of an arc, too. The notion of the length of a curve may be defined on the basis of the general measure theory and it may be proved that for arcs of the form (8) this definition is consistent with the definition given here by means of the integral (7). However, since we do not use here the measure theory, we *define* the length of the arc (8) as equal to the integral (7).

The approximation of the length of an arc by the length of inscribed polygonal lines which we have considered above is the generalization of the approximation of the length of a circumference by the perimeters of the inscribed polygons, known from the elementary geometry. We shall prove that the length of an arc of the circle in the sense of elementary geometry is equal to the length in the sense of the definition assumed here.

EXAMPLE. Let $y = \sqrt{r^2-x^2}$, $r\cos\beta \leqslant x \leqslant r\cos a$ ($0 < a < \beta < \pi$). The considered arc is a part of the semicircle with the radius r and the centre O. Let us calculate the integral (7). We have

$$y' = \frac{-x}{\sqrt{r^2-x^2}}, \quad \text{whence} \quad \int \sqrt{1+(y')^2}\,dx = \int \frac{r\,dx}{\sqrt{r^2-x^2}}.$$

Substituting $x = r\cos t$ we find

$$\int \frac{r\,dx}{\sqrt{r^2-x^2}} = -rt = -r\arccos\frac{x}{r}.$$

Hence

$$\int_{r\cos\beta}^{r\cos a} \sqrt{1+(y')^2}\,dx = (\beta-a)r$$

in accordance with the formula known from elementary geometry.

Let us write

$$s(x) = \int_a^x \sqrt{1+[f'(t)]^2}\,dt.$$

The function $s(x)$, i.e. the length of the arc treated as a function of the abscissa of the right (varying) end of the arc is an increasing function. Thus, there exists a function $x = x(s)$ inverse to $s(x)$, which represents x as a function of the length of the arc. There hold the formulae:

$$(9) \quad \frac{ds}{dx} = \sqrt{1+\left(\frac{dy}{dx}\right)^2} \quad \text{and} \quad \frac{dx}{ds} = \frac{1}{\sqrt{1+\left(\frac{dy}{dx}\right)^2}}.$$

Denoting by a the angle between the tangent to the curve and the X-axis, we have $\frac{dy}{dx} = \tan a$ and so $\frac{dx}{ds}$ $= \cos a$. This means that the derivative is the *direction cosine* of the tangent to the curve $y = f(x)$ with respect to the X-axis.

This follows also from the fact that

$$\frac{dy}{ds} = \frac{dy}{dx}\frac{dx}{ds} = \tan a \cdot \cos a = \sin a \quad \text{and} \quad \left(\frac{dx}{ds}\right)^2 + \left(\frac{dy}{ds}\right)^2 = 1.$$

Remark. The "length element" may be symbolically represented analogously to the Pythagorean formula: $ds^2 = dx^2 + dy^2$, being of a clear geometrical content (cf. the notion of the differential, § 7.13).

The curvature. Let us denote as previously by a the angle between the tangent and the X-axis. This angle is a function of x; considering x as a function of the length s we express a as a function $a(s)$ of the variable s. The derivative $k = \dfrac{da}{ds}$ is called the *curvature* of the

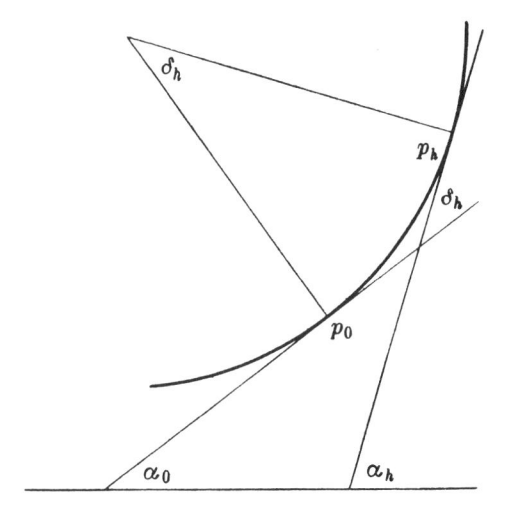

Fig. 23

curve $y = f(x)$ at a given point, the inversion $\dfrac{1}{|k|} = \varrho$ is called the *radius of curvature*.

Geometrically, the curvature k is interpreted as follows:

Let us draw the tangents to the considered curve at the points

$$p_0 = [x(s_0), y(s_0)] \quad \text{and} \quad p_h = [x(s_0+h), y(s_0+h)].$$

The angles between these tangents and the positive direction of the X-axis are $a_0 = a(s_0)$ and $a_h = a(s_0+h)$, respectively. The angle δ_h between these two tangents is $a_h - a_0$. This is the same angle as the angle between the normals to the curve at the points p_0 and p_h. Let us consider the quotient of this angle by the length of the arc $p_0 p_h$, i.e. by h. The curvature of the curve $y = f(x)$ at the point p_0 is the limit of this quotient as h tends to 0.

In particular, if our curve is an arc of the circle $y = \sqrt{r^2 - x^2}$, $-r \leqslant x \leqslant r$, then $h = -r\delta_h$ and so the considered quotient has the constant value $-\dfrac{1}{r}$. Thus the same value has the curvature k, and the radius of the curvature is equal to the radius of the circle.

In the case of a straight line the normals are parallel. Thus $\delta_h = 0$ always and the curvature is equal to 0; hence the radius of the curvature is infinite.

If $k \neq 0$, then the point lying on the normal in the distance ϱ from the point p_0 on the same side of the tangent on which the curve lies locally, is called the *centre of curvature*. We may prove that this point is the limit position to which the point of the intersection of the normals at the points p_0 and p_h tends as h tends to 0.

The curvature k may be expressed in the coordinates x and y as follows. Taking into account that $a = \arctan\left(\dfrac{dy}{dx}\right)$ and basing on formula (9) we obtain

$$\frac{da}{ds} = \frac{da}{dx}\frac{dx}{ds} = \frac{\dfrac{d^2 y}{dx^2}}{1 + \left(\dfrac{dy}{dx}\right)^2} \cdot \frac{1}{\sqrt{1 + \left(\dfrac{dy}{dx}\right)^2}},$$

i.e.

(10) $$k = \frac{\dfrac{d^2 y}{dx^2}}{\left[1 + \left(\dfrac{dy}{dx}\right)^2\right]^{3/2}}.$$

The centre of mass. Let a system of n material points with the coordinates $(x_1, y_1), \ldots, (x_n, y_n)$ and with the masses $m_1, \ldots, m_n$ be given on the plane. The centre of mass of this system is the point with the coordinates

(11) $\quad \bar{x} = \dfrac{m_1 x_1 + \ldots + m_n x_n}{m_1 + \ldots + m_n}, \quad \bar{y} = \dfrac{m_1 y_1 + \ldots + m_n y_n}{m_1 + \ldots + m_n}.$

To define the centre of mass of an arc $y = f(x)$, $a \leqslant x \leqslant b$, with a constant density g we proceed as follows. Let us represent x as a function of the length s, i.e. $x = x(s)$ (as above); hence $y = f[x(s)] = y(s)$. Let S denote the length of the arc $y = f(x)$, $a \leqslant x \leqslant b$. Let us divide this arc into n equal parts. This division constitutes a partition of the interval $0S$ into n equal intervals; let the points of the partition be

$$0 = s_{n,0},\ s_{n,1},\ \ldots,\ s_{n,n} = S.$$

The point with the coordinates $x(s_{n,k})$, $y(s_{n,k})$ is the end of the k-th arc of the n-th partition; the mass of this arc corresponds to this point. This mass is the product of the density by the length of the arc, i.e. it is equal to $g \cdot (s_{n,k} - s_{n,k-1})$. Thus given n, the centre of mass of the system of points

$$[x(s_{n,1}), y(s_{n,1})], \ldots, [x(s_{n,n}), y(s_{n,n})]$$

is the point with the coordinates

$$x_n = \frac{1}{S} \sum_{k=1}^{n} x(s_{n,k})(s_{n,k} - s_{n,k-1}),$$

$$y_n = \frac{1}{S} \sum_{k=1}^{n} y(s_{n,k})(s_{n,k} - s_{n,k-1}),$$

since the mass of the arc is equal to gS.

The limit position of the point (x_n, y_n) as n tends to ∞ is the position of the centre of mass of the considered arc. Thus the coordinates of the centre of mass are

(12) $\quad \bar{x} = \lim\limits_{n=\infty} x_n = \dfrac{1}{S} \int\limits_0^S x\, ds \quad$ and $\quad \bar{y} = \lim\limits_{n=\infty} y_n = \dfrac{1}{S} \int\limits_0^S y\, ds.$

10.6. The volume and surface area of a solid of revolution

Let a continuous function $y = f(x)$, $a \leqslant x \leqslant b$, be given. Moreover, let $f(x) \geqslant 0$. Let us denote as in § 10.4 by P the region constituted by points x, y satisfying the conditions $a \leqslant x \leqslant b$, $0 \leqslant y \leqslant f(x)$. Rotating the region P around the X-axis we obtain a solid of revolution. Let us denote its volume by W. We have

$$(13) \qquad W = \pi \int_a^b y^2 dx.$$

In order to explain this formula, let us consider a normal sequence of partitions (see § 10.3, 2). Let

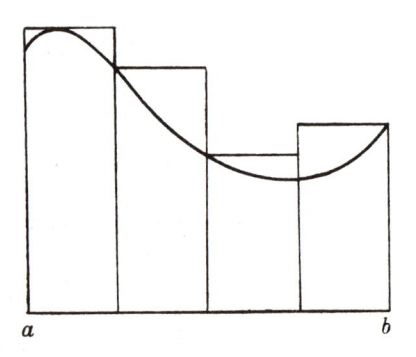

Fig. 24

$f(x_{n,k})$ be the upper bound of the function f in the interval $a_{n,k-1}, a_{n,k}$. Replacing the arc of our curve lying over this interval by the segment $y = f(x_{n,k})$ we obtain by the rotation around the X-axis a cylinder with the base $\pi f(x_{n,k})^2$ and with the height $a_{n,k} - a_{n,k-1}$, and so with the volume

$$\pi f(x_{n,k})^2 (a_{n,k} - a_{n,k-1}).$$

Proceeding in this way for each k (n being constant) we replace the considered solid of revolution by another one, circumscribed on the given solid and made up of a certain number of cylinders. The volume of this solid

10. DEFINITE INTEGRALS 249

is the sum of the volumes of the particular cylinders: denoting this volume by W_n, we have then

$$W_n = \pi \sum_{k=1}^{l_n} f(x_{n,k})^2 (a_{n,k} - a_{n,k-1}).$$

By Theorem 2 of § 10.3 we obtain in the limit

$$W = \lim_{n=\infty} W_n.$$

Analogously, replacing the upper bound by the lower bound in the above construction we prove that W is the limit of the volumes of certain solids inscribed in the considered solid of revolution everyone of which is a sum of a certain number of cylinders.

Let us denote by B the area of the surface obtained by the rotation of the arc $y = f(x)$, $a \leqslant x \leqslant b$, i.e. the area of the lateral surface of the considered solid of revolution. We have the formula

(14) $$B = 2\pi \int_a^b y \sqrt{1 + (y')^2}\, dx$$

(assuming the continuity of the derivative $f'(x)$).

To establish this formula we approximate the considered surface by surfaces the area of which is known from the elementary geometry. For this purpose let us consider a sequence of polygonal lines approximating the arc $y = f(x)$, $a \leqslant x \leqslant b$, defined in § 10.5 (cf. Fig. 10). The surface obtained by the rotation of such a polygonal line around the X-axis constitutes of a finite number of truncated cones. Since the generator of a truncated cone is the segment joining the points $[a_{n,k-1}, f(a_{n,k-1})]$ and $[a_{n,k}, f(a_{n,k})]$, i.e. a segment of the length

$$\sqrt{(a_{n,k} - a_{n,k-1})^2 + [f(a_{n,k}) - f(a_{n,k-1})]^2},$$

and the radii of the bases are of the length $f(a_{n,k})$ and $f(a_{n,k-1})$, respectively, so the area of its lateral surface is equal to

$$\pi[f(a_{n,k-1}) + f(a_{n,k})] \cdot \sqrt{(a_{n,k} - a_{n,k-1})^2 + [f(a_{n,k}) - f(a_{n,k-1})]^2},$$

as follows from a formula known from elementary geometry.

Hence, denoting by B_n the area of the surface obtained by the rotation of the n-th polygonal line, we get

$$B_n = \pi \sum_{k=1}^{l_n} [f(a_{n\,k-1}) + f(a_{n,k})] \cdot$$

$$\cdot \sqrt{(a_{n,k} - a_{n,k-1})^2 + [f(a_{n,k}) - f(a_{n,k-1})]^2} =$$

$$= \pi \sum_{k=1}^{l_n} [f(a_{n,k-1}) + f(a_{n,k})] \cdot \sqrt{1 + [f'(x_{n,k})]^2} (a_{n,k} - a_{n,k-1}),$$

where $a_{n,k-1} \leqslant x_{n,k} \leqslant a_{n,k}$.

We shall prove that

(15) $$B = \lim_{n=\infty} B_n.$$

For this purpose let us compare B_n with

$$C_n = 2\pi \sum_{k=1}^{l_n} f(x_{n,k}) \cdot \sqrt{1 + [f'(x_{n,k})]^2} (a_{n,k} - a_{n,k-1}).$$

By Theorem 2 of § 10.3 with the function $f(x)$ replaced by the product $f(x) \cdot \sqrt{1 + [f'(x)]^2}$ we conclude that $B = \lim_{n=\infty} C_n$. Thus it remains to prove that $\lim_{n=\infty} (B_n - C_n) = 0$.

Now,

$$B_n - C_n = \pi \sum_{k=1}^{l_n} [f(a_{n,k-1}) - f(x_{n,k}) + f(a_{n,k}) - f(x_{n,k})] \cdot$$

$$\cdot \sqrt{1 + [f'(x_{n,k})]^2} (a_{n,k} - a_{n,k-1}).$$

According to the assumption that the considered sequence of partitions of the interval of integration is a normal one we may assume that if the points x and x' belong to the same interval of the n-th partition, then $|f(x) - f(x')| < \varepsilon$ (where ε is a given positive number). Thus

$$|f(a_{n,k-1}) - f(x_{n,k})| < \varepsilon \quad \text{and} \quad |f(a_{n,k}) - f(x_{n,k})| < \varepsilon.$$

10. DEFINITE INTEGRALS

Denoting by M the upper bound of the function $\sqrt{1+[f'(x)]^2}$ in the interval ab we obtain immediately $|B_n - C_n| < 2\pi \varepsilon M (b-a)$. Hence $\lim_{n=\infty}(B_n - C_n) = 0$.

In this way the formula (15) is proved.

A simpler formula for the surface of a solid of revolution is obtained treating x as a function of the length of the arc, similarly as in § 10.5. Namely, applying the first of equalities (9) we have

$$(16) \qquad B = 2\pi \int_a^b y\sqrt{1+(y')^2}\,dx = 2\pi \int_0^S y\,ds,$$

where S denotes the length of the arc $y = f(x)$, $a \leqslant x \leqslant b$.

Formula (16) may be transformed as follows:

$$(17) \qquad B = S \cdot 2\pi \frac{1}{S}\int_0^S y\,ds = S \cdot 2\pi \bar{y},$$

where $\bar{y}$ denotes the ordinate of the centre of mass of the arc $y = f(x)$, $a \leqslant x \leqslant b$ (cf. (12)).

In other words, *the area of the surface obtained by a rotation of the arc $y = f(x)$ around the X-axis is the product of the length of this arc and the path of the centre of mass* (Guldin's theorem).

EXAMPLES. (α) A cylinder. Let $f(x)$ be of a constant value r for $0 \leqslant x \leqslant h$. The region P constituted by the points x, y such that $0 \leqslant x \leqslant h$ and $0 \leqslant y \leqslant r$ is a rectangle. By a full rotation of this rectangle around the X-axis, a cylinder with the height h and with the radius of the base r is obtained. Hence the volume of this cylinder is equal to $\pi r^2 h$. The same result will be obtained applying formula (13), since

$$W = \pi \int_0^h r^2\,dx = \pi r^2 h.$$

The formula (14) gives for the lateral surface of the cylinder (taking into account that $y' = 0$) the formula

$$B = 2\pi \int_0^h r\,dx = 2\pi r h,$$

in accordance with the known formula from geometry.

(β) A c o n e. Let $y = cx$, $0 \leqslant x \leqslant h$. In this case the region P is a triangle. By the rotation, a cone with the height h and with the radius of the base $r = ch$ is obtained. By formula (13) the volume of this cone is equal to

$$\pi \int_0^h c^2 x^2 \, dx = \tfrac{1}{3}\pi c^2 h^3 = \tfrac{1}{3}\pi r^2 h.$$

According to formula (14) the lateral surface of the cone is equal to

$$2\pi \int_0^h cx \sqrt{1+c^2}\, dx = \pi c h^2 \sqrt{1+c^2} = \pi r l,$$

where l denotes the length of the generator (because $l = h\sqrt{1+c^2}$).

(γ) A s p h e r e. Let $y = \sqrt{r^2 - x^2}$, $-r \leqslant x \leqslant r$. The set of points defined by these conditions is a semicircle. Rotating the semicircle around the X-axis we obtain a sphere with the radius r. The volume of this sphere is

$$\pi \int_{-r}^{r} (r^2 - x^2)\, dx = \pi \int_{-r}^{r} r^2\, dx - \pi \int_{-r}^{r} x^2\, dx = 2\pi r^3 - \tfrac{2}{3}\pi r^3 = \tfrac{4}{3}\pi r^3.$$

To evaluate the surface of the sphere, formula (16) may be applied. For this purpose we express y as a function of the length of the arc s, i.e. $y = r\sin\tfrac{s}{r}$. We obtain the following value of the surface of the sphere:

$$2\pi \int_0^{\pi r} r \sin\tfrac{s}{r}\, ds = 2\pi r^2 \int_0^{\pi} \sin t\, dt = 2\pi r^2 [-\cos t]_0^{\pi} = 4\pi r^2.$$

10. DEFINITE INTEGRALS

10.7. Two mean-value theorems

FIRST THEOREM. *Let two continuous functions f and g be given in an interval $a \leqslant x \leqslant b$. Moreover, let the function g be of a constant sign. Then*

$$(18) \qquad \int_a^b f(x)g(x)\,dx = f(\xi) \int_a^b g(x)\,dx,$$

where ξ is a suitable chosen value satisfying the condition $a \leqslant \xi \leqslant b$.

Indeed, let m and M indicate the lower and the upper bound of the function f in the interval $a \leqslant x \leqslant b$, respectively. Then we have

$$m \leqslant f(x) \leqslant M$$

for every x.

Let us assume that $g(x) \geqslant 0$ everywhere. Multiplying the above double inequality by $g(x)$ we obtain then

$$mg(x) \leqslant f(x)g(x) \leqslant Mg(x)$$

and hence, by integration (cf. § 10.2, 7),

$$m \int_a^b g(x)\,dx \leqslant \int_a^b f(x)g(x)\,dx \leqslant M \int_a^b g(x)\,dx.$$

Hence we conclude (assuming the function g to be not identically equal to 0, which can be obviously supposed) that

$$m \leqslant \left\{ \int_a^b f(x)g(x)\,dx : \int_a^b g(x)\,dx \right\} \leqslant M.$$

According to the Darboux property of the function f (which is a consequence of its continuity), this function assumes all values contained between m and M. Therefore there exists a ξ in the interval ab such that

$$f(\xi) = \int_a^b f(x)g(x)\,dx : \int_a^b g(x)\,dx,$$

i.e. that formula (18) is satisfied.

In the case when $g(x) \leqslant 0$ everywhere, the proof is completely analogous.

Remark. Assuming that $g(x) \neq 0$, always we obtain a shorter proof of the above theorem applying the Cauchy theorem stated in § 7.5, (12). Let us denote by $H(x)$ the primitive function of the function $f(x)g(x)$ and by $G(x)$ the primitive function of the function $g(x)$. Then we have
$$H'(x) = f(x)g(x), \qquad G'(x) = g(x),$$
$$\int_a^b f(x)g(x)\,dx = H(b) - H(a), \qquad \int_a^b g(x)\,dx = G(b) - G(a).$$

By the Cauchy theorem,
$$\frac{H(b) - H(a)}{G(b) - G(a)} = \frac{H'(\xi)}{G'(\xi)} = \frac{f(\xi)g(\xi)}{g(\xi)} = f(\xi),$$

and formula (18) follows.

SECOND THEOREM. *If the function f is continuous and the function g is monotone and possesses the second derivative continuous in the interval $a \leqslant x \leqslant b$, then*

(19) $$\int_a^b f(x)g(x)\,dx = g(a)\int_a^\xi f(x)\,dx + g(b)\int_\xi^b f(x)\,dx,$$

where ξ is a suitable chosen value belonging to the interval ab.

Let us denote by $F(x)$ a primitive function of the function $f(x)$, i.e. $F'(x) = f(x)$. Since the derivative of a monotone function is of a constant sign (cf. Theorem 2 of § 7.7), so we may apply the first mean-value theorem to the product of the functions $F(x)$ and $g'(x)$. Hence, applying the rule for integration by parts, we obtain

$$\int_a^b f(x)g(x)\,dx = \int_a^b g(x)F'(x)\,dx = [g(x)F(x)]_a^b - \int_a^b F(x)g'(x)\,dx$$
$$= [g(b)F(b) - g(a)F(a)] - F(\xi)\int_a^b g'(x)\,dx$$
$$= g(b)F(b) - g(a)F(a) - F(\xi)g(b) + F(\xi)g(a),$$

since $\int_a^b g'(x)\,dx = g(b) - g(a)$.

10. DEFINITE INTEGRALS

Consequently,

$$\int_a^b f(x)g(x)\,dx = g(a)[F(\xi)-F(a)]+g(b)[F(b)-F(\xi)],$$

whence we obtain formula (19) taking into account that

$$F(\xi)-F(a) = \int_a^\xi f(x)\,dx \quad \text{and} \quad F(b)-F(\xi) = \int_\xi^b f(x)\,dx.$$

EXAMPLES AND APPLICATIONS. The mean-value theorems are often applied in order to estimate the value of integrals. We shall consider some of their applications.

(α) Let $0 < a < b$. We shall prove that

(20) $$\lim_{n=\infty}\int_a^b \frac{\sin nx}{x}\,dx = 0.$$

For this purpose we shall prove that

(21) $$\left|\int_a^b \frac{\sin cx}{x}\,dx\right| < \frac{4}{ca}, \quad \text{if} \quad c > 0.$$

We apply the second mean-value theorem substituting $f(x) = \sin cx$ and $g(x) = 1/x$. We obtain

$$\int_a^b \frac{\sin cx}{x}\,dx = \frac{1}{a}\int_a^\xi \sin cx\,dx + \frac{1}{b}\int_\xi^b \sin cx\,dx$$

$$= \frac{1}{ac}(\cos ca - \cos c\xi) + \frac{1}{bc}(\cos c\xi - \cos cb).$$

Since $|\cos t| \leqslant 1$ and $a < b$, we conclude that

$$\left|\int_a^b \frac{\sin cx}{x}\,dx\right| \leqslant \frac{2}{ac} + \frac{2}{bc} < \frac{4}{ac}.$$

The formula (21) implies the formula (20) immediately.

Remark. The formula (20) is also an immediate consequence of the formula (γ) in § 10.2.

(β) Applying the first mean-value theorem, the Taylor formula may be deduced in the following way.

We substitute in the formula

$$\int zy^{(n+1)}dx = zy^{(n)} - z'y^{(n-1)} + \ldots + (-1)^n z^{(n)} y + \\ + (-1)^{n+1} \int z^{(n+1)} y \, dx$$

proved in § 9.3, 5, (γ), $z = (b-x)^n$, $y = f(x)$, assuming the function f to possess a continuous $n+1$-th derivative in the interval $a \leqslant x \leqslant b$.

Integrating from a to b we obtain

$$\int_a^b (b-x)^n y^{(n+1)} dx = [(b-x)^n y^{(n)} + n(b-x)^{n-1} y^{(n-1)} + \ldots + n! y]_a^b$$
$$= n! f(b) - \{(b-a)^n f^{(n)}(a) + n(b-a)^{n-1} f^{(n-1)}(a) + \ldots + n! f(a)\}.$$

On the other hand, the function $(b-x)^n$ being of a constant sign in the interval $a \leqslant x \leqslant b$, we obtain by the first mean-value theorem

$$\int_a^b (b-x)^n f^{(n+1)}(x) \, dx$$
$$= f^{(n+1)}(\xi) \cdot \int_a^b (b-x)^n dx = \frac{(b-a)^{(n+1)}}{n+1} \cdot f^{(n+1)}(\xi).$$

Comparing the obtained formulae we get the Taylor formula with the remainder in the Lagrange-form.

10.8. Methods of approximate integrations. Lagrange interpolation

Given $n+1$ points $x_0, x_1, \ldots, x_n$ on the X-axis and $n+1$ values $y_0, y_1, \ldots, y_n$, there may be defined a polynomial $w(x)$ of the n-th degree assuming at these points the values $y_0, y_1, \ldots, y_n$, i.e. such that

(22) $\qquad w(x_k) = y_k \quad$ for $\quad k = 0, 1, \ldots, n$.

First we shall define for each k a polynomial $u_k(x)$ of n-th degree with the property that $u_k(x_k) = 1$ and $u_k(x_m) = 0$ for each $m \neq k$.

Namely, we write

$$u_k(x) = \frac{(x-x_0)\cdot\ldots\cdot(x-x_{k-1})(x-x_{k+1})\cdot\ldots\cdot(x-x_n)}{(x_k-x_0)\cdot\ldots\cdot(x_k-x_{k-1})(x_k-x_{k+1})\cdot\ldots\cdot(x_k-x_n)}.$$

By means of the polynomials $u_k(x)$, we define the polynomial $w(x)$:

(23) $\qquad w(x) = y_0 u_0(x) + y_1 u_1(x) + \ldots + y_n u_n(x).$

We see at once that this is a polynomial of the n-th degree, satisfying the condition (22). This is the so called Lagrange interpolation polynomial.

It is worth while noting that the polynomial $w(x)$ is determined uniquely, i.e. if a polynomial $\overline{w}$ of n-th degree satisfies the condition (22), too, then it is identical with w; for their difference is a polynomial of degree $\leqslant n$ vanishing at $n+1$ points (namely, at the points $x_0, x_1, \ldots, x_n$).

When a function expressing a relation between certain physical quantities is known only in an empirical way, i.e. if only a finite number of values of this function (by means of measurements) is known, then an approximate evaluation of the integral of this function is applied, e.g. the Lagrange interpolation, and then the integral of the obtained polynomial is evaluated.

We obtain an especially simple formula if $n = 2$ and $x_0 = a$, $x_1 = \dfrac{a+b}{2}$, $x_2 = b$, i.e. when we draw a parabola through the given three points. As can be easily calculated, we obtain

(24) $\qquad \displaystyle\int_a^b w(x)\,dx = \frac{b-a}{6}(y_0 + 4y_1 + y_2).$

This is the Simpson formula giving an approximate value of the integral $\int_a^b f(x)\,dx$. The approximation will become more accurate if we divide the interval ab into smaller intervals and if we apply the Simpson formula to everyone of them. Namely, dividing the interval ab

into $2n$ equal intervals by means of points $x_0, x_1, ..., x_{2n}$, where $x_0 = a$, $x_{2n} = b$, we obtain the *general Simpson formula* for the approximate value of an integral:

$$\frac{b-a}{6n}[y_0 + y_{2n} + 2(y_2 + y_4 + ... + y_{2n-2}) + 4(y_1 + y_3 + ... + y_{2n-1})].$$

EXAMPLE. Let us apply the Simpson formula to evaluate $\log 2 = \int_1^2 \frac{dx}{x}$, dividing the interval $1 \leqslant x \leqslant 2$ into halves. We have

$$x_0 = 1, \quad y_0 = 1, \quad x_1 = 1.5, \quad y_1 = 0.667...,$$
$$x_2 = 2, \quad y_2 = 0.5.$$

Thus formula (24) gives the approximate value of $\log 2$:

$$\tfrac{1}{6}(1 + 4 \cdot 0.667 + 0.5) = 0.694...$$

Actually, we have $\log 2 = 0.6931...$

Now we shall mention the so called "trapezoid method" of the approximate evaluation of the integral of a function $f(x) \geqslant 0$ in an interval ab. We divide the interval ab into n equal intervals by the points $x_0 < x_1 < ... < x_n$, where $x_0 = a$, $x_n = b$ and we draw a polygonal line through the points (x_k, y_k), where $k = 0, 1, ..., n$ (cf. Fig. 10). The area s_n between this polygonal line and the X-axis is the sum of areas of n trapezoids $T_1, ..., T_n$, where the area of the trapezoid T_k is equal to $\dfrac{b-a}{n} \cdot \tfrac{1}{2}(y_{k-1} + y_k)$. Thus,

$$s_n = \frac{b-a}{n}\left(\frac{y_0}{2} + y_1 + y_2 + ... + y_{n-1} + \frac{y_n}{2}\right).$$

This is the approximate value of the area between the curve $y = f(x)$, $a \leqslant x \leqslant b$, and the X-axis; speaking otherwise, this is the approximate value of the integral $\int_a^b f(x)dx$.

10.9. Wallis formula [1]

We shall prove the following formula:

(25) $$\pi = \lim_{n=\infty} \frac{1}{n}\left(\frac{2\cdot 4\cdot\ldots\cdot 2n}{1\cdot 3\cdot\ldots\cdot(2n-1)}\right)^2.$$

We shall base our proof on the formulae

$$\int_0^{\pi/2} \sin^{2n}x\,dx = \frac{1\cdot 3\cdot\ldots\cdot(2n-1)}{2\cdot 4\cdot\ldots\cdot 2n}\,\frac{\pi}{2},$$

$$\int_0^{\pi/2} \sin^{2n+1}x\,dx = \frac{2\cdot 4\cdot\ldots\cdot 2n}{3\cdot 5\cdot\ldots\cdot(2n+1)},$$

proved in § 10.1, 5.

From these formulae it follows:

(26) $$\frac{\pi}{2} = \left(\frac{2\cdot 4\cdot\ldots\cdot 2n}{1\cdot 3\cdot\ldots\cdot(2n-1)}\right)^2\cdot\frac{1}{2n+1}\cdot\left\{\int_0^{\pi/2}\sin^{2n}x\,dx : \int_0^{\pi/2}\sin^{2n+1}x\,dx\right\}.$$

We shall prove that

$$\lim_{n=\infty}\left\{\int_0^{\pi/2}\sin^{2n}x\,dx : \int_0^{\pi/2}\sin^{2n+1}x\,dx\right\} = 1.$$

Indeed, in the interval $0 \leqslant x \leqslant \pi/2$ we have the inequalities

$$0 \leqslant \sin^{2n+1}x \leqslant \sin^{2n}x \leqslant \sin^{2n-1}x,$$

whence

$$0 \leqslant \int_0^{\pi/2}\sin^{2n+1}x\,dx \leqslant \int_0^{\pi/2}\sin^{2n}x\,dx \leqslant \int_0^{\pi/2}\sin^{2n-1}x\,dx$$

and so

$$1 \leqslant \int_0^{\pi/2}\sin^{2n}x\,dx : \int_0^{\pi/2}\sin^{2n+1}x\,dx$$

$$\leqslant \int_0^{\pi/2}\sin^{2n-1}x\,dx : \int_0^{\pi/2}\sin^{2n+1}x\,dx \leqslant \frac{2n+1}{2n} = 1 + \frac{1}{2n}.$$

---

[1] John Wallis (1616-1703), an English mathematician.

Thus the considered quotient of integrals tends in the limit to 1.

Hence we conclude by (26) that

(27) $$\frac{\pi}{2} = \lim_{n \to \infty} \left[\left(\frac{2 \cdot 4 \cdot \ldots \cdot 2n}{1 \cdot 3 \cdot \ldots \cdot (2n-1)} \right)^2 \cdot \frac{1}{2n+1} \right].$$

This formula gives formula (25) immediately, since $\lim_{n \to \infty} \frac{n}{2n+1} = \frac{1}{2}$.

Remarks. (α) We apply the Wallis formula also in the following form:

(28) $$\sqrt{\pi} = \lim_{n \to \infty} \frac{(n!)^2 2^{2n}}{\sqrt{n}(2n)!}.$$

Namely, we have

$$2 \cdot 4 \cdot \ldots \cdot 2n = n! \cdot 2^n,$$

$$1 \cdot 3 \cdot \ldots \cdot (2n-1) = \frac{1 \cdot 2 \cdot \ldots \cdot 2n}{2 \cdot 4 \cdot \ldots \cdot 2n} = \frac{(2n)!}{n! \cdot 2^n}.$$

Substituting the above formulae in formula (25) and taking a square root we obtain equality (28).

(β) The Wallis formula may be written also in the form of an infinite product:

(29) $$\frac{\pi}{2} = \prod_{n=1}^{\infty} \frac{4n^2}{4n^2-1} = \frac{2}{1} \cdot \frac{2}{3} \cdot \frac{4}{3} \cdot \frac{4}{5} \cdot \frac{6}{5} \cdot \frac{6}{7} \cdot \ldots$$

Namely, denoting by p_n the partial product

$$p_n = \frac{4 \cdot 1^2}{4 \cdot 1^2 - 1} \cdot \ldots \cdot \frac{4n^2}{4n^2-1} = \left(\frac{2}{1} \cdot \frac{2}{3} \right) \cdot \ldots \cdot \left(\frac{2n}{2n-1} \cdot \frac{2n}{2n+1} \right)$$

$$= \left(\frac{2 \cdot 4 \cdot \ldots \cdot 2n}{1 \cdot 3 \cdot \ldots \cdot (2n-1)} \right)^2 \cdot \frac{1}{2n+1},$$

we have, according to (27), $\lim_{n \to \infty} p_n = \frac{\pi}{2}$. At the same time

$$\lim_{n = \infty} p_n = \prod_{n=1}^{\infty} \frac{4n^2}{4n^2-1},$$

whence the formula (29) follows.

10.10. Stirling formula [1]

We shall deduce from the Wallis formula the following formula due to Stirling:

(30) $$\lim_{n=\infty} \frac{n!}{\sqrt{2\pi n}\, n^n e^{-n}} = 1.$$

Let us write

$$a_n = \frac{n!\, e^n}{n^n \sqrt{n}}.$$

So it is to be proved that $\lim_{n=\infty} a_n = \sqrt{2\pi}$.

First we shall prove the sequence $a_1, a_2, \ldots$ to be convergent. Since it is a sequence of positive terms, it suffices to prove that this sequence is decreasing, i.e. that $\frac{a_n}{a_{n+1}} > 1$. Now,

$$\frac{a_n}{a_{n+1}} = \frac{1}{e}\, \frac{(n+1)^{n+1+1/2}}{n^{n+1/2}(n+1)} = \frac{1}{e}\left(1+\frac{1}{n}\right)^{n+1/2},$$

whence

$$\log \frac{a_n}{a_{n+1}} = \left(n+\frac{1}{2}\right)\log\left(1+\frac{1}{n}\right) - 1.$$

Since (cf. § 7, (54))

$$\left(\frac{1}{x}+\frac{1}{2}\right)\log(1+x) > 1$$

for $0 < x \leqslant 1$, so substituting $x = \frac{1}{n}$ we find

$$\left(n+\frac{1}{2}\right)\log\left(1+\frac{1}{n}\right) > 1, \quad \text{i. e.} \quad \log\frac{a_n}{a_{n+1}} > 0,$$

whence $\frac{a_n}{a_{n+1}} > 1$.

Thus, the sequence $a_1, a_2, \ldots$ is convergent. Let us write $\lim_{n=\infty} a_n = g$. We shall prove that $g \neq 0$.

---

[1] James Stirling (1692-1770)—noted Scottish mathematician.

For this purpose let us note that

$$\log\left(1+\frac{1}{n}\right) = \int_n^{n+1} \frac{dx}{x} < \frac{1}{2}\left(\frac{1}{n}+\frac{1}{n+1}\right),$$

since the arc of the hyperbola $y=\frac{1}{x}$, $n \leqslant x \leqslant n+1$, lies below the segment joining the points $\left(n,\frac{1}{n}\right)$ and $\left(n+1,\frac{1}{n+1}\right)$. Hence

$$\log\frac{a_n}{a_{n+1}} < \frac{1}{4}\left(\frac{1}{n}-\frac{1}{n+1}\right)$$

and so

$$\log\frac{a_1}{a_n} < \frac{1}{4}, \quad \text{i.e.} \quad a_n > e^{3/4}.$$

It remains for us to prove that $g = \sqrt{2\pi}$. To this end we now apply formula (28) taking into account the equations

$$a_n^2 = \frac{(n!)^2 e^{2n}}{n^{2n} n} \quad \text{and} \quad a_{2n} = \frac{(2n)!\, e^{2n}}{(2n)^{2n}\sqrt{2n}}.$$

We obtain

$$\sqrt{\pi} = \lim_{n=\infty} \frac{a_n^2}{a_{2n}\cdot\sqrt{2}} = \frac{g^2}{g\sqrt{2}}, \quad \text{i.e.} \quad g = \sqrt{2\pi}.$$

Remark. The practical importance of the Stirling formula is based on the fact that for large values of n the expression $\sqrt{2\pi n}\, n^n e^{-n}$ gives an approximate value of $n!$. We say that it is an *asymptotic value* of $n!$ (which means that the quotient of these two expressions tends to 1).

10.11*. Riemann integral. Upper and lower Darboux integrals

The theorems of § 10.3 are the starting point for a generalization of the notion of a definite integral to certain classes of discontinuous functions. For this purpose we shall introduce the Darboux integrals.

10. DEFINITE INTEGRALS

Let a bounded (but not necessarily continuous) function $y = f(x)$, $a \leqslant x \leqslant b$, be given. Moreover, let a sequence of partitions of the interval ab into 2^n equal parts ($n = 0, 1, \ldots$) be given. Let us denote the points of the n-th partition by

$$a_{n,0}, a_{n,1}, \ldots, a_{n,2^n}, \quad \text{where} \quad a_{n,0} = a, \quad a_{n,2^n} = b.$$

Let $M_{n,k}$ denote the upper bound of the function f in the interval $a_{n,k-1} a_{n,k}$. Finally, let

$$(31) \qquad S_n = \frac{b-a}{2^n} \sum_{k=1}^{2^n} M_{n,k}.$$

The sequence $\{S_n\}$ is a decreasing sequence (in the wider sense). Indeed, for each $k = 1, 2, \ldots, n$ the intervals $a_{n+1,2k-2} a_{n+1,2k-1}$ and $a_{n+1,2k-1} a_{n+1,2k}$ are contained in the interval $a_{n,k-1} a_{n,k}$. Thus,

$$M_{n+1,2k-1} \leqslant M_{n,k} \quad \text{and} \quad M_{n+1,2k} \leqslant M_{n,k},$$

i.e.

$$\tfrac{1}{2}(M_{n+1,2k-1} + M_{n+1,2k}) \leqslant M_{n,k}.$$

Hence

$$S_{n+1} = \frac{b-a}{2^n}\left(\frac{M_{n+1,0} + M_{n+1,1}}{2} + \ldots + \frac{M_{n+1,2^{n+1}-1} + M_{n+1,2^{n+1}}}{2}\right) \leqslant S_n.$$

Thus the sequence $\{S_n\}$ is non-increasing. It is simultaneously a bounded sequence, since, denoting by m the lower bound of the function f, we have obviously $S_n \geqslant m(b-a)$. Hence it is a convergent sequence. Its limit is called the *upper Darboux integral* and denoted by the symbol

$$(32) \qquad \overline{\int_a^b} f(x)\,dx = \lim_{n=\infty} S_n.$$

Similarly, denoting by $m_{n,k}$ the lower bound of the function f in the interval $a_{n,k-1} a_{n,k}$ and writing

$$(33) \qquad s_n = \frac{b-a}{2^n} \sum_{k=1}^{2^n} m_{n,k}$$

we prove the sequence $\{s_n\}$ to be increasing (in the wider sense) and bounded, and so convergent. Its limit is called the *lower Darboux integral* and is denoted by the symbol

$$(34) \qquad \int\limits_{\underline{a}}^{b} f(x)\,dx = \lim_{n=\infty} s_n\,.$$

If the upper and lower integrals are equal, then the function f is said to be *integrable in the Riemann sense*; the common value of these two integrals is called the *Riemann integral of the function f* and is denoted by $\int_a^b f(x)\,dx$ as in the case of the definite integral of a continuous function. The same symbology may be applied, because (as follows from the theorem 2 of § 10.3 immediately) *the Riemann integral of a continuous function is equal to the definite integral of this function* (in the sense of the definition given in § 10.1).

The functions integrable in the sense of Riemann consist of a larger class of functions than the continuous functions. In particular, bounded functions having a finite number of points of discontinuity (we shall turn to such functions in § 11) and monotone functions are integrable. However, there exist (bounded) functions nonintegrable in the sense of Riemann. An example of such a function is the Dirichlet function which we have already met (cf. § 4, 5), equal to 1 in rational points and to 0 in irrational points; for this function we have always $S_n = 1$ and $s_n = 0$ and so

$$\int\limits_a^{\overline{b}} f(x)\,dx = 1\,, \qquad \int\limits_{\underline{a}}^{b} f(x)\,dx = 0\,.$$

Let us add that there exists a definition of the (Lebesgue) integral which ascribes the integral to a larger class of functions than the Riemann integral ([1]).

---

[1] Cf. S. Saks, *Theory of the Integral*, Monografie Matematyczne, vol. 7 (1937).

10. DEFINITE INTEGRALS

We based the definition of the Darboux integrals on the consideration of a sequence of the "diadic" partitions of the interval of integration. We shall now show that the same result is obtained considering an arbitrary normal sequence of partitions. More strictly, we shall prove the following theorem:

Let a normal sequence of partitions of a segment ab defined by the points

(35) $\quad a'_{n,0} < a'_{n,1} < \ldots < a'_{n,l_n}$, where $a'_{n,0} = a$, $a'_{n,l_n} = b$

be given. Let $M'_{n,k}$ denote the upper bound of the function f in the interval $a'_{n,k-1} a'_{n,k}$ and let

(36) $$S'_n = \sum_{k=1}^{l_n} M'_{n,k}(a'_{n,k} - a'_{n,k-1}).$$

Then

(37) $$\lim_{n=\infty} S'_n = \int_a^{\overline{b}} f(x)\,dx.$$

The proof of this theorem will be based on the following lemma:

Let two partitions of the interval ab,

$a = a_0 < a_1 < \ldots < a_r = b$ and $a = a'_0 < a'_1 < \ldots < a'_{r'} = b$

be given. Let d and d' indicate the length of the greatest interval of the first or of the second partition, respectively. Let M_k and M'_k denote the upper bounds of the function f in the intervals $a_{k-1} a_k$ or $a'_{k-1} a'_k$, respectively, and let

(38) $\quad S = \sum_{k=1}^{r} M_k(a_k - a_{k-1})$ and $S' = \sum_{k=1}^{r'} M'_k(a'_k - a'_{k-1}).$

Finally, let the number M satisfy the inequality $M > |f(x)|$ for every x belonging to the interval ab.

Then the relation

(39) $$S' \leqslant S + 3rMd'$$

holds.

The intervals of the second partition may be classified in $r+1$ classes; a given interval of this partition will belong to the k-th class (for $k = 1, 2, ..., r$), if it is contained in the interval $a_{k-1}a_k$ and to the $r+1$-th class, if it is contained in none of the intervals of the first partition, i.e. if it contains inside one of the points $a_1, a_2, ..., a_{r-1}$. It is clear that some of these classes may be empty.

If the k-th class (for $k \leqslant r$) is non-empty, then let p_k denote the first and j_k the last index i satisfying the condition $a_{k-1} \leqslant a'_i \leqslant a_k$. Since for the intervals $a'_{i-1}a'_i$ of the k-th class there holds the inequality $M'_i \leqslant M_k$, so denoting by $\sum^k$ the sum over all intervals of the k-th class we have

$$\sum^k M'_i(a'_i - a'_{i-1}) \leqslant M_k \sum^k (a'_i - a'_{i-1}) = M_k(a'_{j_k} - a'_{p_k})$$
$$= M_k(a_k - a_{k-1}) - M_k(a'_{p_k} - a_{k-1}) - M_k(a_k - a'_{j_k})$$
$$\leqslant M_k(a_k - a_{k-1}) + M(a'_{p_k} - a_{k-1}) + M(a_k - a'_{j_k})$$
$$\leqslant M_k(a_k - a_{k-1}) + 2Md'.$$

If the k-th class is empty, then let $\sum^k = 0$. Finally let us note that there are at most $r-1$ intervals of the $r+1$-th class. Thus,

$$\sum^{r+1} M'_i(a'_i - a'_{i-1}) \leqslant (r-1)Md'.$$

Consequently,

$$S' = \sum^1 M'_i(a'_i - a'_{i-1}) + ... + \sum^r M'_i(a'_i - a'_{i-1}) +$$
$$+ \sum^{r+1} M'_i(a'_i - a'_{i-1}) \leqslant M_1(a_1 - a_0) + ... +$$
$$+ M_r(a_r - a_{r-1}) + 2rMd' + (r-1)Md',$$

whence the formula (39) follows.

Thus our lemma is proved.

In order to prove the theorem let us consider an $\varepsilon > 0$ and let us choose an index q such that $S_q < \int\limits_a^b f(x)\,dx + \varepsilon$. Since the considered sequence of partitions is normal, we have $d'_n < \dfrac{\varepsilon}{3 \cdot 2^q M}$ for sufficiently large n. Hence, according to the lemma, we have

$$S'_n \leqslant S_q + 3 \cdot 2^q M d'_n < \int\limits_a^{\bar b} f(x)\,dx + 2\varepsilon.$$

Moreover, let us choose to each n an index j_n such that the inequality $\dfrac{b-a}{2^{j_n}} < \dfrac{\varepsilon}{3 l_n M}$ holds. Then

$$S_{j_n} \leqslant S'_n + 3 l_n M \dfrac{b-a}{2^{j_n}} < S'_n + \varepsilon,$$

whence

$$\int\limits_a^{\bar b} f(x)\,dx < S'_n + \varepsilon, \quad \text{because} \quad \int\limits_a^{\bar b} f(x)\,dx \leqslant S_{j_n}.$$

Consequently,

$$\int\limits_a^{\bar b} f(x)\,dx < S'_n + \varepsilon < \int\limits_a^{\bar b} f(x)\,dx + 3\varepsilon.$$

The number ε being arbitrary, this inequality implies formula (37).

Thus, our theorem is completely proved. There is a similar theorem for the lower integral.

The following formula is an important consequence of our theorem:

(40)
$$\int\limits_a^{\bar b} f(x)\,dx = \int\limits_a^{\bar c} f(x)\,dx + \int\limits_c^{\bar b} f(x)\,dx,$$
$$\int\limits_{\underline a}^{b} f(x)\,dx = \int\limits_{\underline a}^{c} f(x)\,dx + \int\limits_{\underline c}^{b} f(x)\,dx,$$

if $a < c < b$.

Indeed, let us divide the segments ac and cb into 2^n equal parts. Let us denote the points of the partition by

(41) $\quad a = a_{n,0} < a_{n,1} < \ldots < a_{n,2^n} = c =$
$\quad\quad\quad\quad = b_{n,0} < b_{n,1} < \ldots < b_{n,2^n} = b.$

Let $M_{n,k}$ denote the upper bound of the function f in the interval $a_{n,k-1}a_{n,k}$ and $P_{n,k}$ the upper bound in the interval $b_{n,k-1}b_{n,k}$. Then we have

$$\int_a^{\bar c} f(x)\,dx = \lim_{n=\infty} \sum_{k=1}^{2^n} M_{n,k}(a_{n,k} - a_{n,k-1})$$

and

$$\int_c^{\bar b} f(x)\,dx = \lim_{n=\infty} \sum_{k=1}^{2^n} P_{n,k}(b_{n,k} - b_{n,k-1}).$$

Moreover,

$$\int_a^{\bar b} f(x)\,dx =$$

$$= \lim_{n=\infty}\Big\{ \sum_{k=1}^{2^n} M_{n,k}(a_{n,k} - a_{n,k-1}) + \sum_{k=1}^{2^n} P_{n,k}(b_{n,k} - b_{n,k-1})\Big\},$$

since the points (41) define a normal sequence of partitions of the segment ab. Hence the first of the formulae (40) follows. The proof of the second one is analogous.

It follows from the formulae (40) that if the function f is integrable in the intervals ac and cb (where $a < c < b$), then it is also integrable in the interval ab and the formula 5 of § 10.2 (on the division of the interval of integration) holds.

Conversely, if a function $f(x)$ is integrable in the interval ab, then it is also integrable in both intervals ac and cb (and so in each interval contained in ab, too).

Indeed, we have by the assumption

$$\int_a^{\bar b} f(x)\,dx = \int_a^b f(x)\,dx = \int_{\underline a}^b f(x)\,dx.$$

Comparing this pair of equations with the formulae (40) we obtain

$$\Big(\int_a^{\bar c} f(x)\,dx - \int_{\underline a}^c f(x)\,dx\Big) + \Big(\int_c^{\bar b} f(x)\,dx - \int_{\underline c}^b f(x)\,dx\Big) = 0.$$

Hence we conclude by the obvious inequalities

$$\int_{\underline{a}}^{c} f(x)\,dx \leqslant \int_{a}^{\bar{c}} f(x)\,dx, \quad \int_{\underline{c}}^{b} f(x)\,dx \leqslant \int_{c}^{\bar{b}} f(x)\,dx$$

that

$$\int_{a}^{\bar{c}} f(x)\,dx = \int_{a}^{c} f(x)\,dx \quad \text{and} \quad \int_{c}^{\bar{b}} f(x)\,dx = \int_{\underline{c}}^{b} f(x)\,dx.$$

Just as for continuous functions we assume the formulae (2) and (3) for arbitrary integrable functions and we prove theorems 1, 2, 6, 7 and 8 of § 10.2. In place of the mean-value theorem we have for arbitrary integrable functions the double inequality

(42) $$m(b-a) \leqslant \int_{a}^{b} f(x)\,dx \leqslant M(b-a),$$

where $a < b$ and where M and m denote the upper bound and the lower bound of the function f in the interval ab respectively.

Theorem 10 of § 10.2 on the differentiability of the function $g(x) = \int_{a}^{x} f(x)\,dx$ does not hold for arbitrary integrable functions. However, it may be proved that the function $g(x)$ is continuous.

Indeed, denoting by M the upper bound of the function $|f(x)|$ in the interval ab we have (by the formulae (40) and (42)):

$$|g(x+h) - g(x)| = \left| \int_{a}^{x+h} f(t)\,dt - \int_{a}^{x} f(t)\,dt \right|$$

$$= \left| \int_{x}^{x+h} f(t)\,dt \right| \leqslant |h| \cdot M,$$

whence $\lim_{h=0} g(x+h) = g(x)$.

Exercises on § 10

1. Prove that $\int_0^{\pi/2} \cos^n x\, dx = \int_0^{\pi/2} \sin^n x\, dx$.

2. Evaluate the sum $\frac{1}{1}\binom{n}{0} - \frac{1}{3}\binom{n}{1} + \frac{1}{5}\binom{n}{2} - \ldots \pm \frac{1}{2n+1}\binom{n}{n}$.

   Hint: apply the substitution $x = \cos t$ to the integral $\int_0^1 (1-x^2)^n\, dx$.

3. Deduce the (Cavalieri) formula
$$\lim_{n=\infty} \frac{1^m + 2^m + \ldots + n^m}{n^{m+1}} = \frac{1}{m+1}$$
applying the integral $\int_0^1 x^m\, dx$ (cf. also Exercises 2 and 3 of § 1).

4. Evaluate $\lim_{n=\infty} n\left(\frac{1}{n^2+1^2} + \ldots + \frac{1}{n^2+n^2}\right)$ (integrating the function $\frac{1}{1+x^2}$).

5. Prove that $\lim_{n=\infty}\left(\frac{1}{n+1} + \frac{1}{n+2} + \ldots + \frac{1}{2n}\right) = \log 2$.

6. Give a geometrical interpretation of the Euler constant (cf. § 7, Example 18) $C = \lim_{n=\infty}\left(1 + \frac{1}{2} + \frac{1}{3} + \ldots + \frac{1}{n} - \log n\right)$, applying the equation $\int_1^n \frac{dx}{x} = \log n$.

7. Prove that $\int_0^\pi \frac{x \sin x}{1 + \cos^2 x}\, dx = \frac{\pi^2}{4}$. Decompose the interval of integration into halves and substitute in the second integral $x = \pi - t$.

8. Prove that $\int_0^{\pi/2} \log \sin x\, dx = -\frac{\pi}{2}\log 2$. (First prove that $\int_0^{\pi/2} \log \sin x\, dx = \int_0^{\pi/2} \log \cos x\, dx = \frac{1}{2}\int_0^{\pi/2} \log \frac{\sin 2x}{2}\, dx$).

9. Find the length of the arc of the parabola given by the equation $y = x^2$, lying between the points with the abscissae 0 and a.

10. Given an ellipse with the excentricity e and with the large axis of length 2, express its perimeter by means of a power series with respect to e.

11. The circle given by the equation $(x-a)^2 + y^2 = r^2$ lying on the XY-plane rotates (in the space) around the Y-axis. Evaluate the area of the surface of the solid obtained in this way, called the *anchor ring* (we assume $a > r$).

12. Evaluate the area of the common part of the circle with the radius r and an ellipse with the axes a and b having the same centre as the circle.

13. Applying the Simpson formula to the integral $\int_0^1 \frac{dx}{1+x^2}$, evaluate the 5 first figures of the decimal expansion of the number $\frac{1}{4}\pi$.

14. Show that the Simpson formula (24) gives the area with an error less than $\dfrac{M(b-a)^4}{576}$, where M denotes the upper bound of $|f'''(x)|$ in the interval $a \leqslant x \leqslant b$.

To prove this, introduce an auxiliary function:

$$g(t) = \int_{c-t}^{c+t} f(x)\,dx - \frac{t}{3}[f(c-t) + 4f(c) + f(c+t)],$$

where $\quad c = \dfrac{a+b}{2},$

and deduce the inequality $-\dfrac{|t|}{3} 2M \leqslant g'''(t) \leqslant \dfrac{|t|}{3} 2M$. Get the required result, integrating this inequality.

15. Prove the Schwarz inequality for integrals:

$$\left(\int_a^b f(x)\,g(x)\,dx\right)^2 \leqslant \int_a^b f^2(x)\,dx \cdot \int_a^b g^2(x)\,dx.$$

(Apply the inequality $\int_a^b [f(x)+c\cdot g(x)]^2 dx \geq 0$ which holds for every c).

16. Give the formula for the radius of curvature at the point x, y for the following curves:

1) hyperbola $xy = 1$,   2) ellipse $\dfrac{x^2}{a^2}+\dfrac{y^2}{b^2} = 1$,

3) parabola $y = 4px^2$,   4) hypocycloid $x^{\frac{2}{3}}+y^{\frac{2}{3}} = a^{\frac{2}{3}}$,

5) the curve $y = x^3$.

17. Find on the exponential curve $y = e^x$ the point at which the curvature is a maximum.

18. Given the hypocycloid from the exercise 16, 4). Evaluate its length, the area of the region contained inside of it, the volume and the area of the solid of revolution obtained by the rotation of this hypocycloid around the X-axis.

19. A system of functions $f_1, \ldots, f_n$ is called *linearly independent*, if no system of n constants $c_1, \ldots, c_n$ exists except of the system of n zeros) such that the equality

$$c_1 f_1(x) + \ldots + c_n f_n(x) = 0$$

holds for every value of x.

Prove that a necessary and sufficient condition for the linear independence of a system of n continuous functions in an interval ab is that the inequality

$$\begin{vmatrix} a_{11} \ldots a_{1n} \\ \cdots \cdots \\ a_{n1} \ldots a_{nn} \end{vmatrix} \neq 0$$

holds, where $a_{km} = \int_a^b f_k(x) f_m(x) dx$.

(Apply the conditions for the existence of non-trivial solutions of a system of n equations

$$a_{k1} x_1 + \ldots + a_{kn} x_n = 0 \quad (k = 1, 2, \ldots, n)$$

with n unknowns).

20. A finite or infinite system of functions $f_1, f_2, \ldots$ is called *orthogonal* in the interval ab, if the conditions

$$\int_a^b f_k(x) f_m(x) dx \begin{cases} = 0 & \text{for } k \neq m, \\ \neq 0 & \text{for } k = m \end{cases}$$

are satisfied.

Give examples of orthogonal systems among the trigonometric functions.

21. Prove that the polynomials (so called *Legendre* [1] *polynomials*) defined by the formula

$$w_n(x) = \frac{1}{2^n n!} \frac{d^n(x^2-1)^n}{dx^n}$$

constitute an orthogonal system in the interval $-1, +1$.
(First prove that if $m < n$, then $\int_{-1}^{+1} w_n(x) \cdot x^m dx = 0$.)

§ 11. IMPROPER INTEGRALS AND THEIR CONNECTION WITH INFINITE SERIES

11.1. Integrals with an unbounded interval of integration

Let a continuous function $y = f(x)$ be given in an infinite interval $x \geqslant a$. Then the integral $\int_a^x f(z) dz$ exists for every $x \geqslant a$. If there exists

$$\lim_{x = \infty} F(x), \quad \text{where} \quad F(x) = \int_a^x f(z) dz,$$

then this limit is denoted by the symbol $\int_a^\infty f(x) dx$ and called the *improper integral* (of the first kind) *of the function f from a to ∞*. Assuming the existence of this

---

[1] Adrien Legendre (1752-1833), a great French mathematician.

limit, we have then

$$(1) \qquad \int_a^\infty f(x)\,dx = \lim_{x=\infty} \int_a^x f(z)\,dz.$$

We say in this case also that the improper integral is *convergent*.

If $\lim\limits_{x=\infty} \int_a^x f(z)\,dz = \infty$ (or $-\infty$), then we write $\int_a^\infty f(x)\,dx = \infty$ (or $-\infty$); in this case we say that the integral is divergent to $+\infty$ or $-\infty$.

Similarly, we define:

$$(2) \qquad \int_{-\infty}^a f(x)\,dx = \lim_{x=-\infty} \int_x^a f(z)\,dz,$$

$$\text{and} \qquad \int_{-\infty}^{+\infty} f(x)\,dx = \int_{-\infty}^0 f(x)\,dx + \int_0^\infty f(x)\,dx.$$

We have by the definition:

$$\int_a^\infty f(x)\,dx = \lim_{x=\infty} [F(x)-F(a)] = \lim_{x=\infty} F(x) - F(a),$$

where F is a primitive function of the function f.

E. g.

$$\int_1^\infty \frac{dx}{x^2} = \left[-\frac{1}{x}\right]_1^\infty = -\lim_{x=\infty}\frac{1}{x} + 1 = 1.$$

Hence this integral is convergent. More generally, for $s > 1$ the integral

$$(3) \qquad \int_1^\infty \frac{dx}{x^s} = \frac{1}{1-s}\left[\frac{1}{x^{s-1}}\right]_1^\infty = \frac{1}{s-1}$$

is convergent.

But for $s = 1$ we obtain an integral divergent to ∞:

$$(4) \qquad \int_1^\infty \frac{dx}{x} = [\log x]_1^\infty = \infty.$$

11. IMPROPER INTEGRALS

An example of an integral having neither a finite nor an infinite limit is $\int_0^\infty \sin x\,dx$, since $\lim\limits_{x=\infty} \cos x$ does not exist.

In the last example the essential fact is that the function under the sign of the integral assumes positive as well as negative values. In fact, the following theorem holds:

1. *If $f(x) \geqslant 0$ for $x \geqslant a$, then the integral $\int_a^\infty f(x)\,dx$ has either a finite or an infinite value (according to whether the function $F(x) = \int_a^x f(z)\,dz$ is bounded or unbounded in the region $x \geqslant a$).*

Indeed, the function $F(x)$ being non-decreasing, it possesses the limit $\lim\limits_{x=\infty} F(x)$ finite or infinite according to whether this function is bounded or unbounded (cf. § 4.7, 1).

Theorem 4′ of § 4.7 implies the following theorem, immediately:

2. *A necessary and sufficient condition for the convergence of the integral $\int_a^\infty f(x)\,dx$ (to a finite limit) is, that to any $\varepsilon > 0$ a number r exists such that the conditions $x > r$ and $x' > r$ imply the inequality*

$$\left| \int_{x'}^x f(z)\,dz \right| < \varepsilon.$$

Indeed, the existence of the limit $\lim\limits_{x=\infty} F(x)$ is equivalent to the condition that to every $\varepsilon > 0$ there exists a number r such that $|F(x) - F(x')| < \varepsilon$, whenever $x > r$ and $x' > r$. Since

$$F(x) - F(x') = \int_{x'}^x f(z)\,dz, \quad \text{so} \quad \left| \int_{x'}^x f(z)\,dz \right| < \varepsilon.$$

Geometrically we interpret the integral $\int_a^\infty f(x)dx$ similarly as the integral in finite limits: if $f(x) \geqslant 0$, then this integral means the area of the unbounded region made up of by points x, y such that $x \geqslant a$ and $0 \leqslant y \leqslant f(x)$.

11.2. Integrals of functions not defined at one point

Let a continuous function f be given in an interval $a \leqslant x < b$ (the interval without the point b). Thus for any x satisfying the above condition the definite integral $\int_a^x f(z)dz$ exists. If there exists the limit $\lim_{x=b-0} F(x)$, where $F(x) = \int_a^x f(z)dz$, then this limit is denoted by the symbol $\int_a^b f(x)dx$, as in the case considered in the previous section, namely, as in the case when the function f is defined and continuous in the whole interval $a \leqslant x \leqslant b$ (together with the ends). In the last case the definition given in § 10 (formula (1)) is consistent with the definition of the symbol $\int_a^b f(x)dx$ given at present. In other words, *if a function $f(x)$ is continuous in the interval $a \leqslant x \leqslant b$, then the integral $\int_a^b f(x)dx$* (in the sense of the definition in § 10, formula (1)) *satisfies the condition*

(5) $\quad \int_a^b f(x)dx = \lim_{x=b-0} F(x), \quad \text{where} \quad F(x) = \int_a^x f(z)dz.$

This equation means that $F(b) = \lim_{x=b-0} F(x)$, i.e. that the function $F(x)$ is (left-side) continuous at the point b; but this follows from Theorem 10 of § 10 (Remark (β)).

The integral of a function f defined in an interval $a \leqslant x < b$ is called an *improper integral* (of the second

11. IMPROPER INTEGRALS

kind). If this integral exists, i.e. if $\lim_{x=b-0} F(x)$ exists, then we say that the integral is *convergent*.

If the function f is defined and continuous in an interval $a < x \leqslant b$, then the integral $\int_a^b f(x)\,dx$ is defined analogously:

$$\int_a^b f(x)\,dx = \lim_{x=a+0} \int_x^b f(z)\,dz.$$

More generally, if a function f is defined and continuous in an open interval $a < x < b$ and if the improper integrals $\int_a^c f(x)\,dx$ and $\int_c^b f(x)\,dx$ are convergent, c being an arbitrary point of the interval $a < x < b$, then by the integral $\int_a^b f(x)\,dx$ we understand

(6) $$\int_a^b f(x)\,dx = \int_a^c f(x)\,dx + \int_c^b f(x)\,dx.$$

It is easily seen that this sum does not depend on the choice of the point c.

Yet more generally, if the function f is defined and continuous in an interval $a \leqslant x \leqslant b$ *except of a finite number of points* (in which the function is not defined or is discontinuous), then by the (improper) integral of the function f in this interval we understand the sum

(7) $$\int_a^b f(x)\,dx = \int_{a_0}^{a_1} f(x)\,dx + \int_{a_1}^{a_2} f(x)\,dx + \ldots + \int_{a_{n-1}}^{a_n} f(x)\,dx,$$

where $a = a_0 < a_1 < a_2 < \ldots < a_n = b$, and all points such that f is undefined or discontinuous at these points are contained in the system of the points $a_0, a_1, a_2, \ldots, a_n$.

So the integral $\int_a^b f(x)\,dx$ is convergent, if all integrals on the right side of formula (7) are convergent.

1. *If a function $f(x)$ is bounded and if it is defined and continuous except of a finite number of points of the interval ab, then the integral $\int_a^b f(x)dx$ is convergent.*

According to the formulae (6) and (7) it is sufficient to prove this theorem in the special case when the function f is defined and continuous for $a \leqslant x < b$.

For this purpose let us note that there exists a necessary and sufficient condition of the convergence of improper integrals of the second kind analogous to the condition given for improper integrals of the first kind (§ 11.1, 2):

2. *A necessary and sufficient condition for the convergence of the integral $\int_a^b f(x)dx$ of a function defined and continuous in $a \leqslant x < b$ is, that to any $\varepsilon > 0$ a number $\delta > 0$ exists such that the conditions $0 < b-x < \delta$ and $0 < b-x' < \delta$ imply the inequality*

$$(8) \qquad \left| \int_{x'}^{x} f(z)dz \right| < \varepsilon.$$

Indeed, writing $F(x) = \int_a^x f(z)dz$ we conclude from Theorem 4 of § 4.7 that the existence of the limit $\lim\limits_{x \to b-0} F(x)$ is equivalent to the inequality $|F(x)-F(x')| < \varepsilon$ (when x and x' satisfy the above mentioned conditions). But

$$F(x)-F(x') = \int_{x'}^{x} f(z)dz,$$

whence the formula (8) follows.

Let us now proceed to the proof of Theorem 1. Given an $\varepsilon > 0$, a number $\delta > 0$ has to be chosen so that the conditions $0 < b-x < \delta$ and $0 < b-x' < \delta$ imply the inequality (8).

The function $f(x)$ is by assumption bounded in $a \leqslant x < b$. Thus, a number M exists such that $M > |f(x)|$

for every x in the considered interval. Let us write $\delta = \varepsilon/M$.
Now,
$$\left|\int_{x'}^{x} f(z)\,dz\right| < |x-x'|M$$
and the inequalities $0 < b-x < \delta$ and $0 < b-x' < \delta$ imply $|x-x'| < \delta = \varepsilon/M$, whence the required formula (8) follows.

In this way Theorem 1 is proved. Thus, the notion of an improper integral makes possible a generalization of the notion of the definite integral to bounded functions having a finite number of points of discontinuity.

EXAMPLES. (α) $\int\limits_0^1 \dfrac{dx}{x^s} = \dfrac{1}{1-s}[x^{1-s}]_0^1 = \dfrac{1}{1-s}$ for $s < 1$.

$\int\limits_0^1 \dfrac{dx}{x} = \infty$.

(β) $\int\limits_0^1 \sin\dfrac{1}{x}\,dx$ exists by Theorem 1, since the function under the sign of the integral is continuous and bounded for $0 < x \leqslant 1$.

(γ) Similarly, the integral $\int\limits_0^1 \dfrac{\sin x}{x}\,dx$ exists. In this case the integral is considered to be *seemingly improper*; although the function under the sign of the integral is not defined at the point 0, it possesses a limit at this point. Assuming this limit to be the value of the integrated function at the point 0 we obtain a definite integral (of a continuous function) in the usual sense.

(δ) Analysing the definition of the (improper) integral of a function f defined and continuous in an open interval $a < x < b$ we see that this integral may be defined as the difference $F(b) - F(a)$, where $F(x)$ is an arbitrary function continuous in the closed interval $a \leqslant x \leqslant b$ and satisfying the condition $F'(x) = f(x)$ in the open interval $a < x < b$ (of course, if such a function F exists).

E. g. the integral $\int\limits_{-r}^{r} \dfrac{r\,dx}{\sqrt{r^2-x^2}}$ which expresses the length of a semicircle with a radius r is an improper integral (the denominator of the integrated function vanishes at the ends of the interval of integration). However, since the primitive function of the integrated function, i.e. $F(x) = -r\arccos(x/r)$ is continuous in the whole interval $-r \leqslant x \leqslant r$, the value of the considered integral is expressed by the difference $F(r) - F(-r) = \pi r$ (just as in the example considered in § 10.5).

Remark. *An improper integral of a function bounded and continuous except of a finite number of points is equal to the Riemann integral of this function.* According to the theorem on the division of the interval of integration, the proof is reduced to the case when f is continuous in the interval $a \leqslant x < b$. Moreover, according to the continuity of the Riemann integral treated as a function of the upper limit of integration (cf. § 10.11) and to the integrability of the function f in the sense of Riemann, the Riemann integral of the function f in the interval ab is equal to $\lim\limits_{x=b-0} \int\limits_{a}^{x} f(x)dx$, i.e. $= \int\limits_{a}^{b} f(x)dx$ (the symbols $\int\limits_{a}^{x}$ and $\int\limits_{a}^{b}$ are applied here in the sense of the definition of § 10.1 and § 11.2).

11.3. Calculation formulae

Let the functions $f(x)$ and $g(x)$ be continuous and defined for $a \leqslant x < b$, where b may also denote ∞. Let us assume that the integrals (improper of the first or of the second kind) $\int\limits_{a}^{b} f(x)dx$ and $\int\limits_{a}^{b} g(x)dx$ are convergent. Then the formulae 1 and 2 of § 10.2 remain valid, for

$$\int\limits_{a}^{b} [f(x) \pm g(x)]dx = \lim\limits_{x=b} \int\limits_{a}^{x} [f(x) \pm g(x)]dx$$

$$= \lim\limits_{x=b} \int\limits_{a}^{x} f(x)dx \pm \lim\limits_{x=b} \int\limits_{a}^{x} g(x)dx$$

$$= \int\limits_{a}^{b} f(x)dx \pm \int\limits_{a}^{b} g(x)dx.$$

11. IMPROPER INTEGRALS

It is proved similarly that the formula 2 of § 10.2 holds.

The formula for the integration by parts (i.e. the formula 3) remains also true assuming that $f(b)g(b)$ denotes $\lim\limits_{x=b} f(x)g(x)$.

Similarly, denoting $\lim\limits_{x=b} f(x)$ by $f(b)$ (where b as well as $f(b)$ may be also $= \infty$) the formula 4 for integration by substitution holds. Namely, assuming the integral $\int_{f(a)}^{f(b)} g(y)\,dy$ to be convergent, we have

$$\int_a^b g[f(x)]f'(x)\,dx = \lim_{x=b} \int_a^x g[f(x)]f'(x)\,dx$$
$$= \lim_{x=b} \int_{f(a)}^{f(x)} g(y)\,dy = \lim_{y=f(b)} \int_{f(a)}^{y} g(y)\,dy = \int_{f(a)}^{f(b)} g(y)\,dy,$$

since the integral treated as a function of the upper limit of integration is continuous (§ 10.2, 10).

Assuming the strict monotony of the function f, the convergence of the integral $\int_a^b g[f(x)]f'(x)\,dx$ implies the convergence of the integral $\int_{f(a)}^{f(b)} g(y)\,dy$. Hence the previously proved relation may be reversed; namely, according to the equation $\lim\limits_{y=f(b)} h(y) = b$, where $h(y)$ denotes the function inverse to $f(x)$, the previous argument may be performed also in the reverse direction.

Other calculation formulae: the formula 5 on the division of the interval of integration and the formulae 6-8 may be also generalized easily to the case of improper integrals.

Remark. Let a function f be *piecewise continuous* in an interval ab. This means (cf. § 5.1) that there exists a system of points $a_0 < a_1 < a_2 < ... < a_n$, where $a_0 = a$ and $a_n = b$, such that the function f_k defined by the con-

ditions $f_k(x) = f(x)$ for $a_{k-1} < x < a_k$, $f_k(a_{k-1}) = f(a_{k-1}+0)$ and $f_k(a_k) = f(a_k-0)$ is continuous in the whole interval $a_{k-1} \leqslant x \leqslant a_k$ ($k = 1, 2, \ldots, n$).
Then we have

$$\int_a^b f(x)\,dx = \sum_{k=1}^n \int_{a_{k-1}}^{a_k} f(x)\,dx = \sum_{k=1}^n \int_{a_{k-1}}^{a_k} f_k(x)\,dx.$$

Using this equation and taking into account the fact that $\int_{a_{k-1}}^{a_k} f_k(x)\,dx$ is a proper integral we state easily that the formula for the integral of a sum remains valid if the functions f and g are piecewise continuous and that the formula for the integration by substitution remains also true assuming the functions $g[f(x)]$ and $f'(x)$ to be piecewise continuous.

11.4. Examples

(α) To evaluate the integral $\int_0^\infty \dfrac{dx}{(1+x^2)^n}$, we substitute $x = \tan t$. Since $\tan 0 = 0$ and $\lim\limits_{t \to \pi/2-0} \tan t = \infty$, we have

$$\int_0^\infty \frac{dx}{(1+x^2)^n} = \int_0^{\pi/2} \cos^{2n} t \frac{d\tan t}{dt}\,dt = \int_0^{\pi/2} \cos^{2n-2} t\,dt$$

$$= \frac{1 \cdot 3 \cdot \ldots \cdot (2n-3)}{2 \cdot 4 \cdot \ldots \cdot (2n-2)} \frac{\pi}{2} \quad \text{for} \quad n > 1 \quad \text{(cf. § 10.1, (ϵ))}.$$

The same integral may be evaluated applying the recurrence formula (§ 9.3, 5, (β)):

$$\int \frac{dx}{(1+x^2)^n} = \frac{1}{2n-2} \cdot \frac{x}{(1+x^2)^{n-1}} + \frac{2n-3}{2n-2} \int \frac{dx}{(1+x^2)^{n-1}}$$

and taking into account that

$$\left[\frac{x}{(1+x^2)^{n-1}}\right]_0^\infty = 0 \quad \text{and} \quad \int_0^\infty \frac{dx}{1+x^2} = [\arctan x]_0^\infty = \pi/2.$$

(β) To prove the convergence of the integral $\int_0^\infty \frac{\sin x}{x} dx$ we shall apply the second mean-value theorem (§ 10.7) taking into account the fact that the function $\frac{1}{x}$ is decreasing. We obtain

$$\int_a^b \frac{\sin x}{x} dx = \frac{1}{a} \int_a^\xi \sin x \, dx + \frac{1}{b} \int_\xi^b \sin x \, dx.$$

Let an $\varepsilon > 0$ be given. An r has to be defined so that the conditions $b > a > r$ imply (cf. § 11.1, 2):

$$\left| \int_a^b \frac{\sin x}{x} dx \right| < \varepsilon.$$

Now,

$$\left| \int_u^v \sin x \, dx \right| \leqslant 2 \quad \text{and so} \quad \left| \int_a^b \frac{\sin x}{x} dx \right| \leqslant 2 \left(\frac{1}{a} + \frac{1}{b} \right) < \frac{4}{a} < \frac{4}{r}.$$

Thus, it suffices to assume $r = \frac{4}{\varepsilon}$.

We note that

(9) $$\lim_{n=\infty} \int_0^a \frac{\sin nx}{x} dx = \int_0^\infty \frac{\sin x}{x} dx \quad (a > 0).$$

Namely, let us substitute $y = nx$. We obtain

(10) $$\int_0^a \frac{\sin nx}{x} dx = \int_0^{na} \frac{\sin y}{y} dy$$

and

$$\lim_{n=\infty} \int_0^{na} \frac{\sin y}{y} dy = \int_0^\infty \frac{\sin y}{y} dy,$$

because $\lim_{n=\infty} na = \infty$.

Remark. Let us write $I_n = \int_{n\pi}^{(n+1)\pi} \frac{\sin x}{x} dx$. Then we have $\int_0^\infty \frac{\sin x}{x} dx = \sum_{n=0}^\infty I_n$. This is an alternating series (which is also easily to illustrate geometrically on the graph of the function $y = \dfrac{\sin x}{x}$, cf. Fig. 25) with (absolutely) decreasing terms tending to 0. Indeed, substituting $y = x - n\pi$ we have

$$I_n = \int_0^\pi \frac{\sin(y+n\pi)}{y+n\pi}\,dy = \pm \int_0^\pi \frac{\sin y}{y+n\pi}\,dy$$

and

$$\int_0^\pi \frac{\sin y}{y+n\pi}\,dy < \int_0^\pi \frac{\sin y}{y+(n-1)\pi}\,dy = \mp I_{n-1}.$$

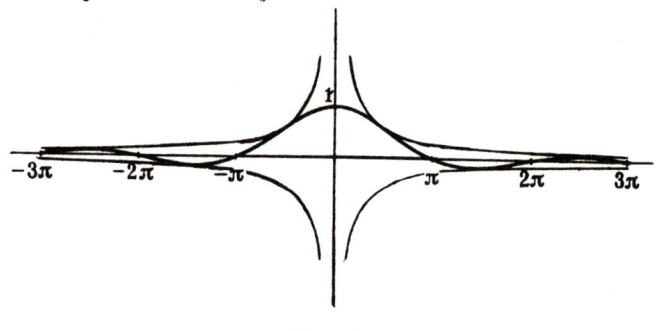

Fig. 25

Hence it follows easily that for every $a > 0$, we have the formula

(11) $$\left| \int_0^a \frac{\sin x}{x}\,dx \right| \leqslant \int_0^\pi \frac{\sin x}{x}\,dx.$$

This formula together with formula (10) gives

(12) $$\left| \int_0^a \frac{\sin nx}{x}\,dx \right| \leqslant \int_0^\pi \frac{\sin x}{x}\,dx.$$

(γ) Similarly as in the previous example, we prove the convergence of the integral $\int_0^\infty \sin(x^2)\,dx$ (the Fresnel integral applied in optics) although we do not know the indefinite integral. Substituting $x^2 = t$ we obtain

$$\int_1^\infty \sin(x^2)\,dx = \frac{1}{2} \int_1^\infty \frac{\sin t}{\sqrt{t}}\,dt.$$

Applying the second mean-value theorem to the last integral we find that

$$\int_a^b \frac{\sin t}{\sqrt{t}}\,dt = \frac{1}{\sqrt{a}} \int_a^\xi \sin t\,dt + \frac{1}{\sqrt{b}} \int_\xi^b \sin t\,dt$$

and we prove the convergence of the integral $\int_1^\infty \frac{\sin t}{\sqrt{t}}\,dt$ by the same arguments as in the previous example. Thus, the integral $\int_1^\infty \sin(x^2)\,dx$ is convergent, too, and so does the integral $\int_0^\infty \sin(x^2)\,dx$, since the integral $\int_0^1 \sin(x^2)\,dx$ is a proper one.

(δ) We have proved previously the convergence of the integral $\int_0^\infty \frac{\sin x}{x}\,dx$. Now, we shall prove that

(13) $$\int_0^\infty \frac{\sin x}{x}\,dx = \frac{\pi}{2}.$$

According to (9),

$$\int_0^\infty \frac{\sin x}{x}\,dx = \lim_{n=\infty} \int_0^{\pi/2} \frac{\sin nx}{x}\,dx,$$

whence

$$\int_0^\infty \frac{\sin x}{x}\,dx = \lim_{n=\infty} \int_0^{\pi/2} \frac{\sin(2n+1)x}{x}\,dx.$$

We shall reduce the last integral to the integral $\int_0^{\pi/2} \frac{\sin(2n+1)x}{\sin x}\,dx$, which we evaluate now. Namely, we shall show that

(14) $$\int_0^{\pi/2} \frac{\sin(2n+1)x}{\sin x}\, dx = \frac{\pi}{2}.$$

Indeed, it follows from the formula (cf. § 1, (2))

$$\frac{1}{2} + \cos t + \cos 2t + \ldots + \cos nt = \frac{\sin \frac{2n+1}{2} t}{2 \sin \frac{t}{2}}$$

that

$$\frac{\sin(2n+1)x}{\sin x} = 1 + 2\cos 2x + 2\cos 4x + \ldots + 2\cos 2nx,$$

whence

$$\int_0^{\pi/2} \frac{\sin(2n+1)x}{\sin x}\, dx =$$

$$= \frac{\pi}{2} + 2\int_0^{\pi/2} \cos 2x\, dx + \ldots + 2\int_0^{\pi/2} \cos 2nx\, dx = \frac{\pi}{2},$$

since all integrals appearing in expression in the middle part of this formula vanish.

Hence the formula (14) is proved. It remains to prove that

$$\lim_{n \to \infty} \int_0^{\pi/2} \left[\frac{\sin(2n+1)x}{\sin x} - \frac{\sin(2n+1)x}{x}\right] dx = 0,$$

i. e. that

$$\lim_{n \to \infty} \int_0^{\pi/2} \frac{x - \sin x}{x \sin x} \sin(2n+1)x\, dx = 0.$$

Now, the last equation is a direct consequence of the formula

$$\lim_{n = \infty} \int_0^{\pi/2} f(x) \sin nx\, dx = 0$$

which we have proved already (§ 10.2, (γ)), after the substitution

$$f(x) = \frac{x - \sin x}{x \sin x} \quad \text{for} \quad x > 0 \text{ and } f(0) = 0.$$

The function f defined in this way is continuous, since $\lim\limits_{x=0} \frac{x - \sin x}{x \sin x} = 0$ (cf. § 8.4, (18)).

(ε) Dirichlet integral: $\int\limits_0^a f(x) \frac{\sin nx}{x} dx$.

We shall prove that *if the function f is monotone and if f' is continuous in the interval $0 \leqslant x \leqslant a$ $(a > 0)$, then*

(15) $$\lim_{n = \infty} \int\limits_0^a f(x) \frac{\sin nx}{x} dx = \frac{\pi}{2} f(0).$$

Let a number $\varepsilon > 0$ be given. We choose c in such a way that

(16) $$|f(c) - f(0)| < \varepsilon \quad \text{and} \quad 0 < c < a.$$

In order to estimate the difference $\int\limits_0^a f(x) \frac{\sin nx}{x} dx - \frac{\pi}{2} f(0)$, let us write

$$\int\limits_0^a f(x) \frac{\sin nx}{x} dx - \frac{\pi}{2} f(0) = \int\limits_0^c [f(x) - f(0)] \frac{\sin nx}{x} dx +$$
$$+ f(0) \left(\int\limits_0^c \frac{\sin nx}{x} dx - \frac{\pi}{2} \right) + \int\limits_c^a f(x) \frac{\sin nx}{x} dx$$

and let us estimate each of the three terms on the right side separately.

To the first term we may apply the second mean-value theorem, since the function $g(x) = f(x) - f(0)$ is monotone. Taking into account that $g(0) = 0$ we obtain

$$\int\limits_0^c [f(x) - f(0)] \frac{\sin nx}{x} dx = [f(c) - f(0)] \int\limits_{\xi_n}^c \frac{\sin nx}{x} dx,$$
$$0 \leqslant \xi_n \leqslant c.$$

By (12),

$$\left| \int_{\xi_n}^{c} \frac{\sin nx}{x} dx \right| \leqslant 2 \int_{0}^{\pi} \frac{\sin x}{x} dx.$$

This inequality together with the inequality (16) gives

(17) $$\left| \int_{0}^{c} [f(x)-f(0)] \frac{\sin nx}{x} dx \right| \leqslant 2\varepsilon \int_{0}^{\pi} \frac{\sin x}{x} dx.$$

To estimate the second term let us note that by (9) and (13),

$$\lim_{n=\infty} \int_{0}^{c} \frac{\sin nx}{x} dx = \frac{\pi}{2}.$$

Hence we have

(18) $$\left| \int_{0}^{c} \frac{\sin nx}{x} dx - \frac{\pi}{2} \right| < \varepsilon$$

for sufficiently large n.

Finally, replacing $f(x)$ in the formula § 10.2, (γ) by $\frac{f(x)}{x}$ and taking into account the inequality $0 < c < a$, we obtain

$$\lim_{n=\infty} \int_{c}^{a} f(x) \frac{\sin nx}{x} dx = 0,$$

whence we conclude that

(19) $$\left| \int_{c}^{a} f(x) \frac{\sin nx}{x} dx \right| < \varepsilon$$

holds for sufficiently large n.

The estimations (17)-(19) give the required estimation for

$$\left| \int_{0}^{a} f(x) \frac{\sin nx}{x} dx - \frac{\pi}{2} f(0) \right|,$$

whence we obtain formula (15).

Remark. The formula (15) remains valid, if we replace the assumption of continuity of f' and that of monotony of f by a weaker assumption, namely, by the assumption that the function f is *piecewise continuous together with its derivative and piecewise monotone*; here $f(0)$ has to be replaced by $f(+0)$ (in the case when the function f is discontinuous at the point 0), i.e.

$$(20) \qquad \lim_{n=\infty} \int_0^a f(x) \frac{\sin nx}{x} dx = \frac{\pi}{2} f(+0).$$

Indeed, our assumption means (cf. § 5.1) that there exists a system of points $0 = a_0 < a_1 < a_2 < \ldots < a_m = a$ such that the function f_k defined by the conditions

$$f_k(x) = f(x) \quad \text{for} \quad a_{k-1} < x < a_k$$

and $\quad f_k(a_{k-1}) = f(a_{k-1}+0), \quad f_k(a_k) = f(a_k-0)$

is continuous together with its derivative and monotone in the whole interval $a_{k-1} \leqslant x \leqslant a_k$, for each $k \leqslant m$.

Applying the theorem on the division of the interval to the improper integrals we obtain

$$\int_0^a f(x) \frac{\sin nx}{x} dx = \int_{a_0}^{a_1} f_1(x) \frac{\sin nx}{x} dx + \ldots + \int_{a_{m-1}}^{a_m} f_m(x) \frac{\sin nx}{x} dx.$$

Hence

$$\lim_{n=\infty} \int_0^a f(x) \frac{\sin nx}{x} dx = \sum_{k=1}^m \lim_{n=\infty} \int_{a_{k-1}}^{a_k} f_k(x) \frac{\sin nx}{x} dx.$$

But this gives the formula (20), since by (15) we have

$$\lim_{n=\infty} \int_{a_0}^{a_1} f_1(x) \frac{\sin nx}{x} dx = \frac{\pi}{2} f_1(0) = \frac{\pi}{2} f(+0),$$

and, on the other hand, we have by § 10.2, (γ),

$$\lim_{n \to \infty} \int_{a_{k-1}}^{a_k} f_k(x) \frac{\sin nx}{x} dx = 0$$

for each $k > 1$, the function $\dfrac{f_k(x)}{x}$ being continuous in the interval $a_{k-1} \leqslant x \leqslant a_k$.

(ζ) The following formula which will be used in the theory of Fourier series follows easily from the formula (15):

$$(21) \quad \lim_{n=\infty} \int_0^a f(x) \frac{\sin nx}{\sin x}\, dx = \frac{\pi}{2} f(+0), \quad 0 < a < \pi;$$

here we assume as previously the function f to be piecewise continuous (together with its derivative) and piecewise monotone.

For this purpose it is enough to note that if we have $0 \leqslant u < v < \pi$, then

$$\lim_{n=\infty}\left\{ \int_u^v f(x) \frac{\sin nx}{\sin x}\, dx - \int_u^v f(x) \frac{\sin nx}{x}\, dx \right\}$$

$$= \lim_{n=\infty} \int_u^v f(x) \frac{x - \sin x}{x \sin x} \sin nx\, dx = 0.$$

For, taking into account the equation $\lim\limits_{x=0} \dfrac{x - \sin x}{x \sin x} = 0$ (cf. § 8.4, (18)) and the inequality $x < \pi$, we may substitute to the function $f(x)$ in the formula § 10.2, (γ), the function $f(x) \dfrac{x - \sin x}{x \sin x}$.

(η) Poisson([1]) integral $\int_0^\infty e^{-x^2}\, dx = \dfrac{\sqrt{\pi}}{2}$.

First of all let us note that applying the substitution $x = z\sqrt{n}$, we obtain

$$(22) \quad \int_0^\infty e^{-x^2}\, dx = \sqrt{n} \int_0^\infty e^{-nz^2}\, dz.$$

([1]) Siméon Poisson, an author of numerous works in the calculus of probability, in mechanics, physics and astronomy.

In order to be able to compare the last integral with the integrals

$$(23) \quad \int_0^1 (1-x^2)^n \, dx = \frac{2 \cdot 4 \cdot \ldots \cdot 2n}{3 \cdot 5 \cdot \ldots \cdot (2n+1)}$$

$$\text{and} \quad \int_0^\infty \frac{dx}{(1+x^2)^n} = \frac{1 \cdot 3 \cdot \ldots \cdot (2n-3)}{2 \cdot 4 \cdot \ldots \cdot (2n-2)} \cdot \frac{\pi}{2}$$

which we already know (cf. § 10.2, (5) and Example (α)), we apply the double inequality

$$1 - x^2 \leqslant e^{-x^2} \leqslant \frac{1}{1+x^2}$$

which follows from the inequality $e^t \geqslant 1+t$ (cf. § 8.4, (16)) by substitution $t = -x^2$ and $t = x^2$. Then we have

$$(1-x^2)^n \leqslant e^{-nx^2} \leqslant \frac{1}{(1+x^2)^n}$$

(the first inequality being valid for $|x| \leqslant 1$), whence

$$\int_0^1 (1-x^2)^n \, dx \leqslant \int_0^1 e^{-nx^2} \, dx \leqslant \int_0^\infty e^{-nx^2} \, dx \leqslant \int_0^\infty \frac{dx}{(1+x^2)^n}.$$

Taking into account the formulae (22) and (23) we conclude that

$$\sqrt{n} \, \frac{2 \cdot 4 \cdot \ldots \cdot 2n}{3 \cdot 5 \cdot \ldots \cdot (2n+1)} \leqslant \int_0^\infty e^{-x^2} \, dx \leqslant \frac{1 \cdot 3 \cdot \ldots \cdot (2n-3)}{2 \cdot 4 \cdot \ldots \cdot (2n-2)} \frac{\pi}{2} \sqrt{n}.$$

Denoting by a_n the left-side term of this relation and by b_n the right-side one, we have $a_n \leqslant \int_0^\infty e^{-x^2} \, dx \leqslant b_n$ for each $n = 1, 2, \ldots$

Now, we have by the Wallis formula (§ 10.9, (25)):

$$\frac{\sqrt{\pi}}{2} = \lim_{n=\infty} c_n \quad \text{where} \quad c_n = \frac{1}{2\sqrt{n}} \cdot \frac{2 \cdot 4 \cdot \ldots \cdot 2n}{1 \cdot 3 \cdot \ldots \cdot (2n-1)}.$$

Hence $a_n = c_n \cdot \dfrac{2n}{2n+1}$ and, since the sequence $\{c_n\}$ is decreasing as easily seen, $\dfrac{\sqrt{\pi}}{2} < c_n$, i. e. $\dfrac{\pi}{2} < 2c_n^2$; hence

$$b_n \leqslant \sqrt{n}\, \frac{1\cdot 3 \cdot \ldots \cdot (2n-3)}{2\cdot 4 \cdot \ldots \cdot (2n-2)}\, 2c_n^2 = c_n \cdot \frac{2n}{2n-1}.$$

Consequently, we conclude that

$$c_n \cdot \frac{2n}{2n+1} \leqslant \int_0^\infty e^{-x^2}\,dx \leqslant c_n \cdot \frac{2n}{2n-1}$$

and since

$$\lim_{n=\infty} \frac{2n}{2n+1} = 1 = \lim_{n=\infty} \frac{2n}{2n-1},$$

we have

$$\int_0^\infty e^{-x^2}\,dx = \lim_{n=\infty} c_n = \frac{\sqrt{\pi}}{2}.$$

11.5. The Gamma function $\Gamma(x) = \int_0^\infty t^{x-1} e^{-t}\,dt$, $x > 0$

This integral possesses for $x < 1$ both singularities: namely, it is an integral in an infinite interval of integration and moreover, the integrated function tends to ∞ as t tends to 0. In order to prove its convergence we shall decompose it into two integrals:

(24) $\qquad \int_0^1 t^{x-1} e^{-t}\,dt \quad \text{and} \quad \int_1^\infty t^{x-1} e^{-t}\,dt$

and we shall prove the convergence of each of these integrals separately.

To prove the convergence of the first of these integrals let us note that $0 < t^{x-1} e^{-t} < t^{x-1}$, since $t > 0$. But the integral $\int_0^1 t^{x-1}\,dt$ is convergent, since $1-x < 1$ (cf. § 11.2, (α));

hence we deduce the convergence of the first of the integrals (24).

Now,

$$t^{x-1}e^{-t} = \frac{t^{x+1}}{e^t} \cdot \frac{1}{t^2} \quad \text{and} \quad \lim_{t=\infty} \frac{t^{x+1}}{e^t} = 0 \quad (\text{cf. § 7 (27)}),$$

whence we have $t^{x-1}e^{-t} < \frac{1}{t^2}$ for sufficiently large t, and the convergence of the second of the integrals (24) follows from the convergence of the integral $\int\limits_1^\infty \frac{dt}{t^2}$ (cf. (3)).

In this way the convergence of the integral $\Gamma(x)$ is proved for every $x > 0$.

Now, we shall prove for positive integers n the formula

(15) $$\Gamma(n) = (n-1)!$$

For this purpose we apply the formula for integration by parts to the integral $\Gamma(x)$ for $x > 1$:

(26) $$\Gamma(x) = -\int\limits_0^\infty t^{x-1} \frac{de^{-t}}{dt} dt$$

$$= -[t^{x-1}e^{-t}]_0^\infty + (x-1)\int\limits_0^\infty t^{x-2}e^{-t} dt = (x-1)\Gamma(x-1).$$

Hence we deduce the formula (25) by induction, applying the equality

$$\Gamma(1) = \int\limits_0^\infty e^{-t} dt = 1 = 0!$$

Let us note that the Poisson integral considered in § 11.4 is one half of the value of the Γ function for $x = \frac{1}{2}$, since substituting $t = z^2$ we find

(27) $$\Gamma\left(\frac{1}{2}\right) = \int\limits_0^\infty \frac{e^{-t}}{\sqrt{t}} dt = 2\int\limits_0^\infty e^{-z^2} dz = \sqrt{\pi}.$$

11.6. The relation between the convergence of an integral and the convergence of an infinite series

THE CAUCHY-MACLAURIN THEOREM. *Let the function $f(x)$, $x \geqslant a$, be continuous, decreasing and positive. Then a necessary and sufficient condition for the convergence of the integral $\int_a^\infty f(x)\,dx$ is the convergence of the infinite series*

$$\sum_{n=1}^{\infty} f(a+n).$$

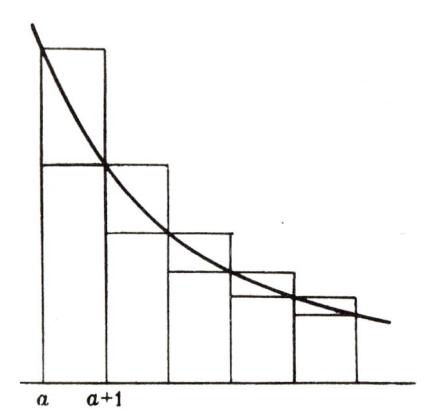

FIG. 26

By the assumption, $f(a+n) \leqslant f(x) \leqslant f(a+n-1)$ for $a+n-1 \leqslant x \leqslant a+n$. Hence

$$\int_{a+n-1}^{a+n} f(a+n)\,dx \leqslant \int_{a+n-1}^{a+n} f(x)\,dx \leqslant \int_{a+n-1}^{a+n} f(a+n-1)\,dx,$$

i.e.

$$f(a+n) \leqslant \int_{a+n-1}^{a+n} f(x)\,dx \leqslant f(a+n-1).$$

Denoting by S_n the partial sum of the series $\sum_{m=1}^{\infty} f(a+m)$, i.e. $S_n = f(a+1) + \ldots + f(a+n)$, we have

$$S_n \leqslant \int_a^{a+n} f(x)\,dx \leqslant S_{n-1} + f(a).$$

11. IMPROPER INTEGRALS

Assuming the integral to be convergent, we conclude that the considered series is bounded: $S_n \leqslant \int_a^\infty f(x)\,dx$, and so it is convergent (as a series with positive terms). Conversely, assuming the convergence of the series, we have

$$\int_a^{a+n} f(x)\,dx \leqslant \sum_{m=0}^\infty f(a+m) \quad \text{for each} \quad n.$$

Hence

$$\int_a^x f(x)\,dx \leqslant \sum_{m=0}^\infty f(a+m) \quad \text{for every} \quad x,$$

since, denoting for a given x by n a positive integer such that $x < a+n$, we have

$$\int_a^x f(x)\,dx \leqslant \int_a^{a+n} f(x)\,dx \leqslant \sum_{m=0}^\infty f(a+m).$$

So the function $\int_a^x f(x)\,dx$ is bounded and, consequently the integral $\int_a^\infty f(x)\,dx$ is convergent (cf. § 11.1, 1).

Remarks. (α) The assumption that the function f is positive may be omitted in the above theorem. Namely, if a decreasing function is negative, then the integral as well as the considered series are divergent to $-\infty$.

(β) If the integral $\int_a^\infty f(x)\,dx$ of a decreasing function is convergent, then $\lim_{x\to\infty} f(x) = 0$, as is easily seen. However, if the function is not decreasing, then the integral may be convergent although this inequality does not hold. The evidence gives here the integral $\int_0^\infty \sin(x^2)\,dx$ which we have considered already in § 11.4, (γ); this integral is convergent but the limit of $\sin(x^2)$ as x tends to ∞ does not exist.

Assuming $f(x) \geq 0$, an even stronger singularity may be obtained. Namely, let us construct an infinite sequence of isosceles triangles with height 1 and having the intervals $(0, 1)$, $(,1\ 1\frac{1}{2})$, ..., $\left(n, n + \dfrac{1}{2^n}\right)$, ... of the X-axis as bases, successively. Let $y = f(x)$ be the graph of an infinite polygonal line constituted by the sides of these

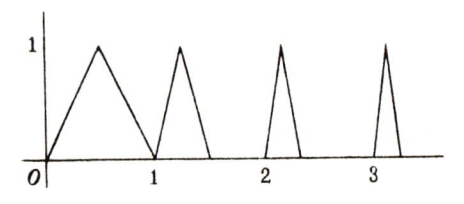

FIG. 27

triangles (different from the bases) and the segments of the X-axis joining the successive triangles. Thus, the region contained between this polygonal line and the X-axis is constituted by the above triangles; hence its area is the sum of their areas, i.e.

$$\int_0^\infty f(x)\,dx = \frac{1}{2} + \frac{1}{4} + \ldots + \frac{1}{2^{n+1}} + \ldots = 1.$$

However, $\lim\limits_{x \to \infty} f(x)$ does not exist.

EXAMPLES. (α) The series $\zeta(s) = \sum\limits_{n=1}^\infty \dfrac{1}{n^s}$ is convergent for every $s > 1$ ([1]), since the integral $\int\limits_1^\infty \dfrac{dx}{x^s}$ is convergent (cf. § 11.1, (3)).

(β) The series $\sum\limits_{n=2}^\infty \dfrac{1}{n(\log n)^s}$ is convergent for $s > 1$ and divergent for $s = 1$.

([1]) The (Riemann) function $\zeta(s)$ is of great importance in the theory of numbers.

To prove this, let us consider the integral $\int_2^\infty \dfrac{dx}{x(\log x)^s}$.
Now,
$$\int \frac{dx}{x(\log x)^s} = \int \frac{d\log x}{(\log x)^s} = \frac{1}{1-s} \cdot \frac{1}{(\log x)^{s-1}} \quad \text{for} \quad s > 1.$$
Since
$$\lim_{x=\infty} (\log x)^{s-1} = \infty,$$
we have
$$\int_2^\infty \frac{dx}{x(\log x)^s} = \frac{1}{(s-1)(\log 2)^{s-1}}.$$

Hence we conclude by the Cauchy-Maclaurin theorem that our series is convergent for $s > 1$. Yet for $s = 1$ this series is divergent, for
$$\int \frac{dx}{x \log x} = \int \frac{d\log x}{\log x} = \log(\log x).$$

According to the equation $\lim_{x=\infty} \log(\log x) = \infty$, this implies the divergence of the integral $\int_2^\infty \dfrac{dx}{x \log x}$.

(γ) Similarly, it may be proved that the integral
$$\int_a^\infty \frac{dx}{x(\log x)[\log(\log x)]^s}$$
is convergent for $s > 1$ and divergent for $s = 1$ (where a is sufficiently large). Hence the series
$$\sum_{n=k}^\infty \frac{1}{n(\log n)[\log(\log n)]^s}$$
is convergent for $s > 1$ and divergent for $s = 1$.

More general results may be obtained considering the product of n successive iterations of the logarithm the last of which is raised to the power s (in the previous example $n = 2$, the zero-iteration $= x$).

This leads to further types of convergent series and to an infinite sequence of infinite series

$$\sum_{n=1}^{\infty}\frac{1}{n}, \quad \sum_{n=2}^{\infty}\frac{1}{n\log n}, \quad \sum_{n=k}^{\infty}\frac{1}{n\log n \log(\log n)} \quad \text{etc.}$$

divergent more and more slowly.

(δ) **Logarithmic criteria.** The comparison of the terms of a given series $\sum_{n=2}^{\infty} a_n$ with the terms of the series $\sum_{n=2}^{\infty}\frac{1}{n(\log n)^s}$ leads to the following convergence criterion: *if $a_n > 0$ and if*

(28) $$\lim_{n=\infty}\frac{\log(na_n)}{\log(\log n)} < -1,$$

then the series $\sum_{n=2}^{\infty} a_n$ is convergent; if this limit is > -1, then the series is divergent.

Indeed, let us assume the inequality (28) to be satisfied and let us denote by $-s$ a number greater than the considered limit and less than -1. Then we have $s > 1$ and

$$\log(na_n) < -s\log(\log n) = \log(\log n)^{-s}$$

for sufficiently large n, whence

$$na_n < (\log n)^{-s}, \quad \text{i.e.} \quad a_n < \frac{1}{n(\log n)^s}.$$

Thus, by the theorem on the comparison of series with positive terms, the convergence of the series $\sum_{n=2}^{\infty}\frac{1}{n(\log n)^s}$ implies the convergence of the series $\sum_{n=2}^{\infty} a_n$.

Yet if the limit considered in the formula (28) is greater than -1, then for sufficiently large n we have

$\log(na_n) > -\log(\log n)$, whence $na_n > (\log n)^{-1}$, i.e. $a_n >$
$> \dfrac{1}{n \log n}$ and the divergence of the series $\sum\limits_{n=2}^{\infty} \dfrac{1}{n \log n}$
implies the divergence of the series $\sum\limits_{n=2}^{\infty} a_n$.

The case when the considered limit is equal to -1 is for our logarithmic criterion a "doubtful" case, i.e. we are not able to state by this criterion whether the series is convergent or divergent. However, in this case stronger logarithmic criteria of the considered in Example (γ), may be applied.

Let us add that there exist series (with positive terms) which do not react to any logarithmic criterion.

11.7. Fourier [1] series

Suppose that we are given a convergent trigonometric series of the form

(29) $\quad f(x) = \tfrac{1}{2} a_0 + (a_1 \cos x + b_1 \sin x) + (a_2 \cos 2x + b_2 \sin 2x) +$
$\qquad + \ldots + (a_n \cos nx + b_n \sin nx) + \ldots$

We note that if the given series is uniformly convergent in the interval $-\pi \leqslant x \leqslant \pi$, then the coefficients a_n and b_n may be expressed easily by means of the sum of the series, i.e. of the function f. For, according to the uniform convergence we have (cf. § 10.2, 12)

$$\int_{-\pi}^{\pi} f(x)\, dx = \int_{-\pi}^{\pi} \frac{a_0}{2}\, dx + \sum_{n=1}^{\infty} \int_{-\pi}^{\pi} (a_n \cos nx + b_n \sin nx)\, dx$$

and since

$$\int_{-\pi}^{\pi} \cos nx\, dx = 0 = \int_{-\pi}^{\pi} \sin nx\, dx \quad \text{and} \quad \int_{-\pi}^{\pi} \frac{a_0}{2}\, dx = \pi a_0,$$

---

[1] Jean Fourier (1768-1830). Fourier introduced the expansions of functions known under his name in connection with his work in the theory of the heat conduction.

we have

(30) $$a_0 = \frac{1}{\pi} \int_{-\pi}^{\pi} f(x)\,dx.$$

To evaluate a_n, we multiply both sides of the identity (29) by $\cos nx$ and we find

$$\int_{-\pi}^{\pi} f(x)\cos nx\,dx = \int_{-\pi}^{\pi} \frac{a_0}{2} \cos nx\,dx +$$
$$+ \sum_{m=1}^{\infty} \int_{-\pi}^{\pi} (a_m \cos mx \cos nx + b_m \sin mx \cos nx)\,dx.$$

just as in the previous calculation.

Since (cf. § 10.1, (δ))

$$\int_{-\pi}^{\pi} \cos^2 nx\,dx = \pi$$

and $\int_{-\pi}^{\pi} \cos mx \cos nx\,dx = 0 = \int_{-\pi}^{\pi} \sin mx \cos nx\,dx$

for $m \neq n$, we have

(31) $$a_n = \frac{1}{\pi} \int_{-\pi}^{\pi} f(x) \cos nx\,dx.$$

Similarly we find

(32) $$b_n = \frac{1}{\pi} \int_{-\pi}^{\pi} f(x) \sin nx\,dx.$$

The following question arises: which functions f may be represented in the form (29), the coefficients a_n and b_n satisfying the formulae (30)-(32) (the so-called *Euler-Fourier formulae*). If such an expansion exists, then it is called the *Fourier series of the function f*.

Since the functions cos and sin are periodic, it is evident that the periodicity of the function f has to be assumed, too. Moreover, we shall assume that the function f is piecewise monotone (cf. § 4, 2).

We shall prove the following

THEOREM. *Every periodic function f with the period 2π (i.e. $f(x+2\pi) = f(x)$), piecewise continuous (together with its derivative), piecewise monotone and satisfying (at the points of discontinuity) the condition*

$$(33) \qquad f(x) = \frac{f(x-0)+f(x+0)}{2}$$

may be expanded in a Fourier series.

Let us denote by $S_n(x)$ the n-th partial sum of the series (29), i.e.

$$(34) \quad S_n(x) = \tfrac{1}{2} a_0 + (a_1 \cos x + b_1 \sin x) + \ldots + $$
$$+ (a_n \cos nx + b_n \sin nx).$$

We have to prove that if the coefficients satisfy the conditions (30)-(32), then

$$(35) \qquad f(x) = \lim_{n=\infty} S_n(x).$$

The above mentioned conditions make it possible for us to transform formula (34) as follows:

$$(36) \quad \pi S_n(x) = \int_{-\pi}^{\pi} \tfrac{1}{2} f(t)\, dt + \int_{-\pi}^{\pi} f(t)(\cos t \cos x + \sin t \sin x)\, dt +$$
$$+ \ldots + \int_{-\pi}^{\pi} f(t)(\cos nt \cos nx + \sin nt \sin nx)\, dt$$
$$= \int_{-\pi}^{\pi} f(t)[\tfrac{1}{2} + \cos(t-x) + \ldots + \cos n(t-x)]\, dt$$
$$= \int_{-\pi}^{\pi} f(t) \frac{\sin(2n+1)\dfrac{t-x}{2}}{2\sin\dfrac{t-x}{2}}\, dt$$

by the known formula (cf. § 1, (2)):

$$\tfrac{1}{2}+\cos t+\cos 2t+\ldots+\cos nt=\frac{\sin\frac{2n+1}{2}t}{2\sin\frac{t}{2}}.$$

Let us substitute $\frac{t-x}{2}=z$. We obtain

$$(37)\qquad \pi S_n(x)=\int\limits_{-\frac{\pi-x}{2}}^{\frac{\pi-x}{2}} f(x+2z)\frac{\sin(2n+1)z}{\sin z}\,dz.$$

Decomposing this integral into two integrals in the intervals from 0 to $\frac{\pi-x}{2}$ and from $\frac{-\pi-x}{2}$ to 0 and substituting in the second integral $z=-y$, we obtain

$$(38)\qquad \pi S_n(x)=\int\limits_0^{\frac{\pi-x}{2}} f(x+2z)\frac{\sin(2n+1)z}{\sin z}\,dz+$$

$$+\int\limits_0^{\frac{\pi+x}{2}} f(x-2y)\frac{\sin(2n+1)y}{\sin y}\,dy.$$

Now, we shall prove the equation (35) for $-\pi<x<\pi$. According to (33) it is sufficient to prove that

$$(39)\qquad \lim_{n=\infty}\int\limits_0^{\frac{\pi-x}{2}} f(x+2z)\frac{\sin(2n+1)z}{\sin z}\,dz=\frac{\pi}{2}f(x+0)$$

and

$$(40)\qquad \lim_{n=\infty}\int\limits_0^{\frac{\pi+x}{2}} f(x-2y)\frac{\sin(2n+1)y}{\sin y}\,dy=\frac{\pi}{2}f(x-0).$$

Given x, let us write $f(x+2z)=g(z)$. Then the function $g(z)$ is piecewise continuous (together with its derivative)

11. IMPROPER INTEGRALS

and piecewise monotone. Hence it satisfies (cf. (21)) the formula

$$(41) \quad \lim_{n=\infty} \int_0^a g(z) \frac{\sin nz}{\sin z} dz = \frac{\pi}{2} g(+0), \quad \text{if} \quad 0 < a < \pi.$$

In this formula a may be replaced by $\frac{\pi-x}{2}$, because the inequality $-\pi < x < \pi$ implies the inequality $0 < \frac{\pi-x}{2} < \pi$. Since $g(+0) = f(x+0)$, formula (41) gives (39) immediately (we restrict ourselves from the sequence $n = 1, 2, \ldots$ to the subsequence of the odd numbers).

Formula (40) is obtained in an exactly similar way, writing $f(x-2y) = h(y)$ and taking into account that $0 < \frac{\pi+x}{2} < \pi$ and that $h(+0) = f(x-0)$.

Thus, our theorem is proved for $-\pi < x < \pi$. It remains to consider the case when $x = -\pi$ (or when $x = \pi$).

Let $x = -\pi$. According to (37) we have

$$\pi S_n(-\pi) = \int_0^\pi f(-\pi+2z) \frac{\sin(2n+1)z}{\sin z} dz.$$

Decomposing this integral into two integrals in the intervals from 0 to $\frac{1}{2}\pi$ and from $\frac{1}{2}\pi$ to π and substituting $y = \pi - z$ in the second integral we obtain

$$\pi S_n(-\pi) = \int_0^{\pi/2} f(-\pi+2z) \frac{\sin(2n+1)z}{\sin z} dz +$$
$$+ \int_0^{\pi/2} f(\pi-2y) \frac{\sin(2n+1)y}{\sin y} dy.$$

The first of these integrals tends to $\frac{1}{2}\pi f(-\pi+0)$ as previously and the second one tends to $\frac{1}{2}\pi f(\pi-0)$, i.e. to $\frac{1}{2}\pi f(-\pi-0)$ (since the function f is periodic).

Hence the formula (35) holds also for $x = -\pi$; hence, according to the periodicity, it holds for every x.

In this way our theorem is proved completely.

11.8. Applications and examples

(α) Let $f(x)$ denote a periodic function with the period 2π given by the following conditions:

$$f(x) = -\tfrac{1}{4}\pi \quad \text{for} \quad -\pi < x < 0, \; f(0) = 0,$$
$$f(x) = \tfrac{1}{4}\pi \quad \text{for} \quad 0 < x < \pi, \; f(\pm\pi) = 0.$$

The graph of this function consists of segments of the length π lying on the straight lines $y = \tfrac{1}{4}\pi$ and $y = -\tfrac{1}{4}\pi$, alternately, and of points on the X-axis being integer multiples of the number π (the ends of the segments do not belong to the graph).

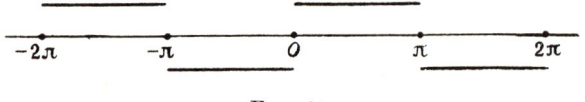

FIG. 28

It is immediately seen that the function defined in this way is piecewise continuous and piecewise monotone (namely, it is monotone in intervals of the form $(m-1)\pi < x < m\pi$, where m is an integer); moreover, the condition (33) is satisfied. The theorem on the expansion in Fourier series may therefore be applied to this function.

We calculate the coefficients a_n and b_n according to formulae (30)-(32).

We obtain

$$a_0 = \frac{1}{\pi} \int_{-\pi}^{0} -\frac{\pi}{4} dx + \frac{1}{\pi} \int_{0}^{\pi} \frac{\pi}{4} dx = 0,$$

$$a_n = \frac{1}{4}\left\{ \int_{-\pi}^{0} -\cos nx\, dx + \int_{0}^{\pi} \cos nx\, dx \right\} = 0 \quad (n > 0),$$

$$b_n = \frac{1}{4}\left\{ \int_{-\pi}^{0} -\sin nx\, dx + \int_{0}^{\pi} \sin nx\, dx \right\} = \frac{1}{4n}(2 - 2\cos n\pi),$$

and so $b_n = 0$ for n even and $b_n = 1/n$ for n odd.

Consequently,

(42) $$f(x) = \frac{\sin x}{1} + \frac{\sin 3x}{3} + \frac{\sin 5x}{5} + \ldots$$

So, for x satisfying the condition $0 < x < \pi$, $f(x)$ may be replaced by $\frac{1}{4}\pi$.

In particular, substituting $x = \frac{1}{2}\pi$ we obtain the Leibniz expansion

$$\frac{\pi}{4} = 1 - \frac{1}{3} + \frac{1}{5} - \frac{1}{7} + \ldots$$

which we already derived (cf. § 7, (51)).

(β) Let $f(x)$ be the function with the period 2π defined as follows:

$f(x) = \frac{1}{2}x$, when $-\pi < x < \pi$, $f(-\pi) = 0 = f(\pi)$.

We see easily that the function f defined in this way is piecewise continuous and piecewise monotone (cf. Fig. 29).

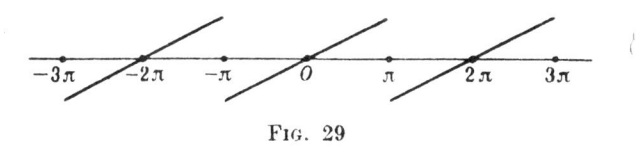

Fig. 29

Moreover, the condition (33) is satisfied. Thus, the function f may be expanded in a Fourier series.

Calculating the coefficients a_n and b_n according to the formulae (30)-(32) we find

(43) $$\frac{x}{2} = \frac{\sin x}{1} - \frac{\sin 2x}{2} + \frac{\sin 3x}{3} - \ldots, \quad -\pi < x < \pi.$$

Let us note that substituting $x = \frac{1}{2}\pi$, we obtain the Leibniz expansion for $\frac{1}{4}\pi$ (just as in the previous example).

(γ) Let us expand the function $f(x) = |x|$ in the interval $-\pi \leqslant x \leqslant \pi$. For other x the function is defined by the condition $f(x + 2m\pi) = f(x)$. Hence it is a continuous function on the whole X-axis. The graph of this function is a polygonal line.

We now calculate the coefficients of the expansion in the Fourier series:

$$a_0 = \frac{1}{\pi} \int_{-\pi}^{\pi} |x|\,dx = \frac{2}{\pi} \int_0^{\pi} x\,dx = \pi,$$

$$a_n = \frac{1}{\pi} \int_{-\pi}^{\pi} |x|\cos nx\,dx = \frac{2}{\pi} \int_0^{\pi} x\cos nx\,dx = 0 \quad \text{or} \quad -\frac{4}{\pi n^2}$$

according as n is even (>0) or odd. For,

$$\int x\cos nx\,dx = \frac{1}{n} x\sin nx - \frac{1}{n} \int \sin nx\,dx$$

$$= \frac{1}{n} x\sin nx + \frac{1}{n^2} \cos nx.$$

Similarly, we find that $b_n = \frac{1}{\pi} \int_{-\pi}^{\pi} |x|\sin nx\,dx = 0$. Thus,

(44) $$|x| = \frac{\pi}{2} - \frac{4}{\pi}\left(\frac{\cos x}{1^2} + \frac{\cos 3x}{3^2} + \frac{\cos 5x}{5^2} + \ldots\right).$$

Let us note that by substituting $x = 0$ we obtain

(45) $$\frac{\pi^2}{8} = \frac{1}{1^2} + \frac{1}{3^2} + \frac{1}{5^2} + \ldots$$

Hence we obtain the Euler formula

(46) $$\frac{\pi^2}{6} = \frac{1}{1^2} + \frac{1}{2^2} + \frac{1}{3^2} + \ldots$$

For, denoting by s the sum of this series, we have

$$\frac{s}{4} = \frac{1}{2^2} + \frac{1}{4^2} + \frac{1}{6^2} + \ldots \quad \text{and so} \quad s - \frac{s}{4} = \frac{1}{1^2} + \frac{1}{3^2} + \frac{1}{5^2} + \ldots,$$

i.e. $\frac{3}{4}s = \frac{1}{8}\pi^2$, that is $s = \frac{1}{6}\pi^2$.

(δ) Let $y = \cos tx$, $-\pi \leqslant x \leqslant \pi$ and let us assume that t is not an integer. The formulae (30)-(32) easily give:

$$a_0 = \frac{2}{\pi t}\sin \pi t, \quad a_n = \frac{2}{\pi}(-1)^n \frac{t}{t^2 - n^2}\sin \pi t, \quad b_n = 0.$$

11. IMPROPER INTEGRALS 307

Thus,

(47) $\quad \pi \cos tx = 2t \sin \pi t \left(\dfrac{1}{2t^2} - \dfrac{\cos x}{t^2 - 1^2} + \dfrac{\cos 2x}{t^2 - 2^2} - \ldots \right).$

In particular, substituting $x = \pi$ and dividing both sides of identity (47) by $\sin \pi t$ we obtain:

(48) $\quad \pi \cot \pi t = \dfrac{1}{t} + 2t \left(\dfrac{1}{t^2 - 1^2} + \dfrac{1}{t^2 - 2^2} + \ldots \right).$

The formula (48) has the following interesting application.

Let us write

$$S(t) = \sum_{n=1}^{\infty} \dfrac{2t}{n^2 - t^2}.$$

Given a number x satisfying the condition $0 < x < 1$, the series $S(t)$ is uniformly convergent in the interval $0 \leqslant t \leqslant x$, since $\dfrac{t}{n^2 - t^2} \leqslant \dfrac{1}{n^2 - x^2}$ and the series $\sum\limits_{n=1}^{\infty} \dfrac{1}{n^2 - x^2}$ is convergent, as is easily seen.

Therefore the series $S(t)$ may be integrated term by term in the interval $0 \leqslant t \leqslant x$. Since

$$\int \dfrac{2t}{n^2 - t^2} dt = -\log(n^2 - t^2),$$

we have $\quad \displaystyle\int_0^x \dfrac{2t}{n^2 - t^2} dt = -\log\left(1 - \dfrac{x^2}{n^2}\right).$

Hence (cf. § 10.2, 12 and § 5, (12)):

$$\int_0^x S(t) dt = -\sum_{n=1}^{\infty} \log\left(1 - \dfrac{x^2}{n^2}\right) = -\log \prod_{n=1}^{\infty} \left(1 - \dfrac{x^2}{n^2}\right).$$

On the other hand,

$$\pi \int \left(\cot \pi t - \dfrac{1}{\pi t}\right) dt = \log \sin \pi t - \log \pi t = \log \dfrac{\sin \pi t}{\pi t}.$$

Thus,

$$\pi \int_0^x \left(\cot \pi t - \frac{1}{\pi t}\right) dt$$

$$= \log \frac{\sin \pi x}{\pi x} - \lim_{t = +0} \log \frac{\sin \pi t}{\pi t} = \log \frac{\sin \pi x}{\pi x},$$

since

$$\lim_{t=+0} \log \frac{\sin \pi t}{\pi t} = \log \lim_{t=+0} \frac{\sin \pi t}{\pi t} = \log 1 = 0.$$

Taking into account the fact that the relation (48) holds for every t satisfying the inequality $0 < t \leqslant x$ and that the function $\cot \pi t - \frac{1}{\pi t}$ has a right-side limit at the point 0, namely $\lim_{t=+0}\left(\cot t - \frac{1}{t}\right) = 0$ (cf. § 8, (19)), we conclude that

$$\pi \int_0^x \left(\cot \pi t - \frac{1}{\pi t}\right) dt = -\int_0^x S(t) \, dt,$$

i.e.

$$\log \frac{\sin \pi x}{\pi x} = \log \prod_{n=1}^{\infty} \left(1 - \frac{x^2}{n^2}\right),$$

therefore

(49) $$\sin \pi x = \pi x \prod_{n=1}^{\infty} \left(1 - \frac{x^2}{n^2}\right), \quad -1 < x < 1.$$

Let us note that by substituting in the formula (49) $x = \frac{1}{2}$, we obtain the Wallis formula already derived (cf. § 10, (29)):

(50) $$\frac{\pi}{2} = \prod_{n=1}^{\infty} \frac{4n^2}{4n^2-1} = \frac{2}{1} \cdot \frac{2}{3} \cdot \frac{4}{3} \cdot \frac{4}{5} \cdot \frac{6}{5} \cdot \frac{6}{7} \cdots = \frac{4}{3} \cdot \frac{16}{15} \cdot \frac{36}{35} \cdots$$

Substituting $x = \frac{1}{4}$, we obtain

$$\frac{\sqrt{2}}{2} = \sin\frac{\pi}{4} = \frac{\pi}{4} \cdot \prod_{n=1}^{\infty}\left[1 - \frac{1}{(4n)^2}\right] = \frac{\pi}{4} \cdot \frac{3\cdot 5}{4\cdot 4}\cdot\frac{7\cdot 9}{8\cdot 8}\cdot\frac{11\cdot 13}{12\cdot 12}\cdots,$$

from which we conclude by (50) that

$$\sqrt{2} = \frac{2 \cdot 2}{1 \cdot 3} \cdot \frac{6 \cdot 6}{5 \cdot 7} \cdot \frac{10 \cdot 10}{9 \cdot 11} \cdot \ldots = \frac{4}{3} \cdot \frac{36}{35} \cdot \frac{100}{99} \cdot \ldots$$

Exercises on § 11

1. Evaluate $\int_0^\infty \frac{dx}{x^2+4}$, $\int_0^\infty \frac{\arctan x}{1+x^2} dx$.

2. Investigate the convergence of the integrals

$$\int_0^1 \frac{\cos x}{\sqrt{x}} dx, \quad \int_0^\infty \frac{\cos x}{1+x^2} dx.$$

3. Let $\lim_{x=\infty} f(x) = 0$ and let a number $c > 0$ be given. Let us write

$$a_n = \int_{a+c(n-1)}^{a+cn} f(x) dx.$$

If the series $a_1 + a_2 + \ldots$ is convergent, then the integral $\int_a^\infty f(x) dx$ is convergent, too, and is equal to $\sum_{n=1}^\infty a_n$.

Apply the above theorem to the proof of the convergence of the integral

$$\int_0^\infty \frac{\sin x}{x} dx$$

4. An integral $\int_a^\infty f(x) dx$ is called *absolutely convergent*, if the integral $\int_a^\infty |f(x)| dx$ is convergent. Prove that an absolutely convergent integral is also convergent in the usual sense.

5. Prove that the series $\sum_{n=k}^\infty \frac{1}{n^2 - a^2}$ is convergent ($k > a$).

6. Prove that $\int_{-1}^{1} \frac{\log(1+x)}{x} dx = \frac{\pi^2}{4}$.

(Expand the function under the sign of the integral in an infinite series and apply the formula (45))

7. Prove that $\frac{1}{1^2} - \frac{1}{2^2} + \frac{1}{3^2} - \frac{1}{4^2} + \ldots = \frac{\pi^2}{12}$ (cf. the formula (45)).

8. Expand the functions:

    1) x^2,   2) $x\cos x$,   3) $|\sin x|$,

    4) $\sinh tx$,   5) $\cosh tx$

in Fourier series.

9. Assuming that the system of functions $f_1, f_2, \ldots$ is orthogonal and that the series $f(x) = \sum_{n=1}^{\infty} a_n f_n(x)$ is uniformly convergent (in the considered interval ab), calculate the coefficients a_n (by means of the functions $f, f_1, f_2, \ldots$).

SUPPLEMENT *

ADDITIONAL EXERCISES

Exercises on § 1

1.1. Prove that
$$|a\sin x + b\cos x| \leqslant \sqrt{a^2+b^2}.$$

1.2. Prove that
$$\frac{|a_1+a_2+\ldots+a_n|}{1+|a_1+a_2+\ldots+a_n|} \leqslant \frac{|a_1|}{1+|a_1|} + \frac{|a_2|}{1+|a_2|} + \ldots + \frac{|a_n|}{1+|a_n|}.$$

1.3. Prove the inequalities
$$\sqrt[m]{a+b} \leqslant \sqrt[m]{a} + \sqrt[m]{b} \quad \text{and} \quad \left|\sqrt[m]{a} - \sqrt[m]{b}\right| \leqslant \sqrt[m]{|a-b|}.$$
for $a \geqslant 0$, $b \geqslant 0$ (where m is a positive integer).

1.4. Prove that if $c \geqslant 1$, then the inequality
$$\left|\frac{a}{1+a^2} - \frac{b}{1+b^2}\right| \leqslant c|a-b|$$
holds for any a and b. Show that the numbers $c < 1$ do not possess this property.

1.5. Prove that
$$\sqrt[m]{1+a} \leqslant 1 + \frac{a}{m}$$
for $a \geqslant -1$ (m is a positive integer).

---
* Written by W. Kołodziej.

1.6. Show that for any positive integers n and k the following inequality holds:

$$2^n > \left(\frac{n}{k}\right)^k.$$

Hint: Substitute a suitable integer m in the inequality $2^m \geq 1 + m$ (obtained from the Bernoulli inequality).

1.7. Find the upper bound and the lower bound of the set of all decimal fractions of the form $0.11\ldots1$.

1.8. Similarly: find the bounds of the set of all numbers of the form $(n+m)^2/2^{nm}$, where n and m are positive integers.

1.9. Let Z_x and Z_y be two non-empty bounded sets of real numbers. Denoting by Z_{x+y} the set of all sums $x+y$, where x belongs to Z_x and y belongs to Z_y, show that

$$\sup Z_{x+y} = \sup Z_x + \sup Z_y,$$
$$\inf Z_{x+y} = \inf Z_x + \inf Z_y,$$

(the symbols $\sup Z$ and $\inf Z$ mean the upper bound and the lower bound, respectively, of the set Z).

Exercises on § 2

2.1. Evaluate the limits

$$\lim_{n=\infty} \left(\sqrt{n+\sqrt{n}} - \sqrt{n}\right), \quad \lim_{n=\infty} \sqrt[n]{1 + 2^{(-1)^n}},$$

$$\lim_{n\to\infty} \sqrt[n]{1^k + 2^k + \ldots + n^k} \quad (k \text{ an integer}), \quad \lim_{n\to\infty}\left(1 + \frac{1}{n^2}\right)^n.$$

2.2. Prove that the sequences

$$\left\{\left(\cos\frac{2\pi n}{3}\right)^n\right\} \quad \text{and} \quad \{(-1)^{n(n+1)/2} \sqrt[n]{n}\}$$

are divergent.

2.3. Similarly: prove the divergence of the sequences

$$\{\sin nt\} \quad \text{and} \quad \{\cos nt\} \quad (0 < t < \pi).$$

Hint: Every interval of the length t contains a number x of the form $x = nt$, where n is an integer.

2.4. Let $\{a_n\}$ be a non-decreasing sequence of positive numbers. Prove that
$$\lim_{n=\infty} \sqrt[n]{a_1^n + a_2^n + \ldots + a_n^n} = \lim_{n=\infty} a_n .$$

2.5. Let a non-empty and bounded-above set Z be given. Prove that the number M is the upper bound of the set Z if and only if $M \geqslant x$ for every x belonging to Z and $\lim_{n=\infty} x_n = M$ for a sequence $\{x_n\}$ with terms belonging to Z.

Express and prove an analogous theorem concerning the lower bound.

2.6. Given a number $c > 0$ we define the sequence $\{a_n\}$ by induction as follows:
$$a_1 > 0, \qquad a_{n+1} = \left(a_n + \frac{c}{a_n}\right).$$
Show that $\lim_{n=\infty} a_n = \sqrt{c}$.

Hint: Prove that the sequence $a_2, a_3, \ldots$ is decreasing.

2.7. Prove that every convergent sequence either contains a greatest term or contains a least term.

2.8. A sequence $\{a_n\}$ is said to be *of bounded variation*, if the sequence $\{s_n\}$ with terms
$$s_n = |a_2 - a_1| + |a_3 - a_2| + \ldots + |a_{n+1} - a_n|$$
is bounded. Prove that every sequence of bounded variation is convergent. Show by an example that the converse theorem is not true.

2.9. Find the limit of the sequence $\{a_n\}$ defined by induction thus:
$$a_1 > 0, \qquad a_{n+1} = \frac{1}{1 + a_n} .$$
Hint: Note that $|a_{n+1} - a_n| \leqslant (\tfrac{1}{2})^{n-1} |a_2 - a_1|$.

2.10. Let $\{a_n\}$ be a sequence bounded neither above nor below and such that $\lim_{n=\infty}(a_{n+1} - a_n) = 0$. Prove that every real number is the limit of a certain subsequence of this sequence.

Exercises on § 3

3.1. Prove the convergence of the series

$$\sum_{n=1}^{\infty} \frac{n[2+(-1)^n]^n}{4^n}, \quad \sum_{n=1}^{\infty} \frac{1\cdot 3\cdot\ldots\cdot(2n-1)}{2\cdot 4\cdot\ldots\cdot 2n}\cdot\frac{1}{2n+1},$$

$$\sum_{n=1}^{\infty} \frac{(-1)^{n-1}}{n+(-1)^{n-1}}, \quad \sum_{n=1}^{\infty} \frac{1}{\sqrt{n}}\sin\frac{\pi n}{8}.$$

3.2. Prove the convergence of the series $\sum_{n=1}^{\infty} \frac{1}{2^{\sqrt{n}}}$.

3.3. Examine the convergence of the series

$$1+\frac{1}{2}-\frac{1}{3}+\frac{1}{4}+\frac{1}{5}-\frac{1}{6}+\ldots$$

3.4. Assuming that t is not an integral multiple of the number 2π, prove that the series $\sum_{n=1}^{\infty} \frac{\cos nt}{n^a}$ is absolutely convergent for $a > 1$ and conditionally convergent for $0 < a \leqslant 1$.

Hint: To prove the divergence of the series $\sum_{n=1}^{\infty} \frac{|\cos nt|}{n^a}$ for $0 < a \leqslant 1$, apply the inequality $|\cos nt| \geqslant \cos^2 nt$.

3.5. Prove that if $a_n \geqslant 0$, then the convergence of the series $\sum_{n=1}^{\infty} a_n$ implies the convergence of the series $\sum_{n=1}^{\infty} a_n^2$.

3.6. Prove that if the series $\sum_{n=1}^{\infty} a_n$ and $\sum_{n=1}^{\infty} b_n$ are convergent and $a_n \leqslant c_n \leqslant b_n$, then the series $\sum_{n=1}^{\infty} c_n$ is convergent, too.

3.7. Let $\sum_{n=1}^{\infty} a_n$ be a divergent series with positive terms. Prove that the series

$$\sum_{n=1}^{\infty} \frac{a_n}{a_1+a_2+\ldots+a_n}$$

is also divergent.

Hint: Apply the Cauchy theorem (§ 3.2.1).

3.8. Prove the convergence of the sequence $\{a_n\}$ with the general term
$$a_n = 2\sqrt{n} - \left(\frac{1}{\sqrt{1}} + \frac{1}{\sqrt{2}} + \ldots + \frac{1}{\sqrt{n}}\right).$$

Hint: Note that every sequence is the sequence of partial sums of a series.

3.9. Prove that the series $\sum_{n=1}^{\infty} (-1)^{[\sqrt{n}]} \frac{1}{n}$ is convergent.

3.10. Prove that the series $\sum_{n=1}^{\infty} (-1)^{[\sqrt{n}]} \frac{1}{\sqrt[4]{n}}$ has the following property: for every real number x one may group its terms in such a way to obtain a series convergent to x.

Hint: Apply the theorem of Exercise 2.10.

Exercises on § 4

4.1. Evaluate the limits
$$\lim_{x=0} \frac{1-\cos x}{x^2}, \quad \lim_{x=0} x \log(x+x^2),$$
$$\lim_{x=0} \frac{\sqrt{x}-1}{x-1}, \quad \lim_{x=0} x \left[\frac{1}{x}\right].$$

4.2. Evaluate the limits
$$\lim_{x=\infty}(\sin \sqrt{x+1} - \sin \sqrt{x}), \quad \lim_{x=\infty} \frac{\log(x+2^x)}{\log(x+3^x)}.$$

4.3. Investigate the existence of the limits
$$\lim_{x=0} \frac{\sqrt{1-\cos x}}{\sin x}, \quad \lim_{x=0}(1+x)^{1/x^2},$$
$$\lim_{x=0}\left(\left[\frac{1}{x}\right] - \frac{1}{x}\right), \quad \lim_{x=\infty} \frac{e^x}{x} \sin \sqrt{x}.$$

4.4 Show by an example that the following theorem is false: if $\lim_{x=a} f(x) = A$ and $\lim_{y=A} g(y) = B$, then $\lim_{x=a} g[f(x)] = B$ (cf. § 4.6 (16)).

4.5. Let

$$f(x) = \sum_{n=1}^{\infty} \frac{1}{n^2 x + 1}.$$

Prove that $\lim\limits_{x=+0} f(x) = +\infty$, $\lim\limits_{x=\infty} f(x) = 0$.

Exercises on § 5

5.1. Evaluate the limits

$$\lim_{x=0} \frac{\sqrt{\cos x} - 1}{x^2}, \quad \lim_{x=0} \frac{\log(1+x)}{x},$$

$$\lim_{x=0} \frac{a^x - 1}{x} \quad (a > 0), \quad \lim_{x=0} (\cos x)^{1/x^2}.$$

5.2. Prove that for every x the following formula holds:

$$e^x = \lim_{n \to \infty} \left(1 + \frac{x}{n}\right)^n.$$

5.3. Suppose that the function f satisfies for any x and y the condition

$$f(x+y) = f(x) + f(y).$$

Prove that if f is continuous at a point a, then it is continuous at every point.

5.4. Let the function f be continuous in the interval $a < x \leqslant b$ and let for every sequence $\{x_n\}$ with rational terms belonging to this interval the condition $\lim\limits_{n=\infty} x_n = a$ imply $\lim\limits_{n=\infty} f(x_n) = g$. Prove that there exists $\lim\limits_{n=\infty} f(x) = g$.

5.5. Assuming that the function f is continuous in the interval $a \leqslant t \leqslant b$, denote by $F(x)$ the upper bound of this function in the interval $a \leqslant t \leqslant x$. Prove that the function F is continuous in the interval $a \leqslant x \leqslant b$.

Hint: If $x_1 < x_2$, then obviously $F(x_1) \leqslant F(x_2)$. Prove that if $F(x_1) < F(x_2)$, then $F(x_2) = f(t_0)$ for a point t_0 of the interval $x_1 \leqslant t \leqslant x_2$; conclude finally that $F(x_2) - F(x_1) \leqslant f(t_0) - f(x_1)$.

5.6. Let a continuous function f be given in an infinite interval $x \geqslant a$. Prove that the existence of the finite $\lim_{x=\infty} f(x)$ implies the boundedness of this function.

5.7. Investigate which of the following functions are uniformly continuous:
$$\frac{1}{\sqrt{x}}, \; e^x, \; x^2, \; \frac{x}{|x|+1}.$$

5.8. Prove that the product of two uniformly continuous and bounded functions is uniformly continuous. Show by considering the example of function $x\sin x$ that the assumption that both functions are bounded is essential.

5.9. Prove that every function f uniformly continuous in a finite interval $a < x < b$ possesses the finite limits $\lim_{x=a} f(x)$ and $\lim_{x=b} f(x)$.

5.10. Let f be a continuous function defined in an infinite interval $x \geqslant a$. Prove that the existence of finite $\lim_{x=\infty} f(x)$ implies the uniform continuity of this function.

5.11. Prove that a function continuous in an infinite interval $x \geqslant a$ and bounded neither above nor below assumes every real value infinitely many times.

5.12. Give an example of a function $y = f(x)$ defined in the interval $0 \leqslant x \leqslant 1$, on-to-one and continuous at the point $x = 0$ and such that the inverse function $x = g(y)$ is discontinuous at the point $y_0 = f(0)$.

Exercises on § 6

6.1. Investigate the uniform convergence of the following sequences of functions:
$$f_n(x) = nx(1-x)^n \quad (0 \leqslant x \leqslant 1),$$
$$f_n(x) = \sqrt[n]{1+x^n} \quad (x \geqslant 1).$$

6.2. Similarly: examine the uniform convergence of the following series of functions:

$$\sum_{n=1}^{\infty} \frac{1}{1+(x-n)^2}, \quad \sum_{n=1}^{\infty} \frac{x^2}{n^4+x^4},$$

$$\sum_{n=1}^{\infty} \frac{(-1)^n}{x+n} \quad (x \geqslant 0), \quad \sum_{n=1}^{\infty} \frac{x^n}{n} \quad (0 \leqslant x < 1).$$

6.3. Prove that the function f given by the formula

$$f(x) = \sum_{n=1}^{\infty} \frac{1}{x^2-n^2}$$

is continuous at every point at which it is defined (i.e. for $x \neq \pm n$).

6.4. Prove that the function

$$f(x) = \sum_{n=1}^{\infty} \sqrt{x}\, e^{-n^2 x}$$

is continuous in the interval $x > 0$ and discontinuous at the point $x = 0$.

6.5. Prove that if the functions $f_1, f_2, \ldots, f_n, \ldots$ are continuous and positive in a closed interval $a \leqslant x \leqslant b$ and the series $\sum_{n=1}^{\infty} f_n(x)$ is convergent to a continuous function, then the convergence is uniform.

(Note that the uniform convergence of the series $\sum_{n=1}^{\infty} f_n(x)$ is equivalent to the uniform convergence of the sequence $R_m(x) = \sum_{n=m+1}^{\infty} f_n(x)$ to the function equal to 0. Assuming that the convergence is not uniform, conclude the existence of a convergent sequence $\{x_m\}$ such that $R_m(x_m) \geqslant \varepsilon > 0$ for infinitely many indices m. Writing $x_0 = \lim_{m=\infty} x_m$ prove that $R_n(x_0) \geqslant \varepsilon$ for every n.)

6.6. Find the radius of convergence of the following power series:

$$\sum_{n=0}^{\infty} [2+(-1)^n]^n x^n, \quad \sum_{n=1}^{\infty} \frac{\varepsilon_n}{n} x^n \text{ (where } \varepsilon_n = 0 \text{ or } 1).$$

6.7. Let r be the radius of convergence of a given power series $\sum_{n=0}^{\infty} a_n x^n$. Prove that if r is finite and $\sum_{n=0}^{\infty} a_n r^n = \infty$, then $\lim_{x=r-0} \sum_{n=0}^{\infty} a_n x^n = \infty$ (a complement to the Abel theorem).

Hint: Reduce the proof to the special case $r = 1$; note also that $\sum_{n=0}^{\infty} a_n x^n = (1-x) \sum_{n=0}^{\infty} s_n x^n$, where $s_n = a_0 + a_1 + \ldots + a_n$.

6.8. Prove that if $a_n \geqslant 0$, then the Abel theorem may be reversed. Speaking more strictly, if r denotes the radius of convergence of the power series $\sum_{n=0}^{\infty} a_n x^n$, then (assuming that r is finite) the existence of the finite $\lim_{x=r-0} \sum_{n=0}^{\infty} a_n x^n = g$ implies the convergence of the series $\sum_{n=0}^{\infty} a_n r^n$ (to g).

6.9. Let $\sum_{n=0}^{\infty} a_n$ and $\sum_{n=0}^{\infty} b_n$ be two convergent series. Prove that if the Cauchy product of these series, i.e. the series $\sum_{n=0}^{\infty} c_n$ with $c_n = a_0 b_n + a_1 b_{n-1} + \ldots + a_n b_0$, is convergent, then

$$\sum_{n=0}^{\infty} c_n = \sum_{n=0}^{\infty} a_n \cdot \sum_{n=0}^{\infty} b_n$$

(compare this with the Cauchy theorem of § 3.8).

Hint: Note that $\sum_{n=0}^{\infty} c_n x^n = \sum_{n=0}^{\infty} a_n x^n \cdot \sum_{n=0}^{\infty} b_n x^n$ for $0 \leqslant x < 1$ and apply the Abel theorem.

6.10. Let r denote the radius of convergence of a given power series $\sum_{n=0}^{\infty} a_n x^n$. Prove that the infinite product $\prod_{n=0}^{\infty}(1+a_n x^n)$ is convergent if $|x| < r$ and divergent if $|x| > r$.

Exercises on § 7

7.1. Prove that if the function f is differentiable in an infinite interval $x > a$ and $\lim_{x=\infty} f(x) = 0$, then the equation $f'(x) = 0$ has in this interval no less roots than the equation $f(x) = 0$.

7.2. Let the function f be differentiable in an interval $a < x < b$ (finite or infinite). Prove that if the derivative f' is bounded, then f is uniformly continuous in this interval.

7.3. Prove that if the function f is differentiable in an infinite interval $x > a$ and $\lim_{x=\infty} f'(x) = \infty$, then it is not uniformly continuous in this interval.

7.4. Let f be a function differentiable in an infinite interval $x > a$. Prove that if $\lim_{x=\infty} f'(x) = g$, then $\lim_{x=\infty} \frac{f(x)}{x} = g$, too.

Hint: One may apply the theorem of exercise 5, § 4.

7.5. Prove that

$$\arctan x + \arctan \frac{1-x}{1+x} = \frac{\pi}{4} \quad (x > -1).$$

Hint: Find the derivative of the left side.

7.6. Prove that if $a > 1$, then

$$\frac{1}{2^{a-1}} \leqslant x^a + (1-x)^a \leqslant 1$$

for $0 \leqslant x \leqslant 1$.

7.7. Prove that if $0 < a < 1$, then

$$(x+y)^a < x^a + y^a$$

for any positive x and y (a generalization of the inequality given in exercise 1.3).

7.8. Prove that the series $\sum_{n=1}^{\infty} x^2 e^{-n^2 x}$ is uniformly convergent in the interval $x \geqslant 0$.

Hint: Find the upper bound of the function $x^2 e^{-n^2 x}$.

7.9. Prove that
$$\sum_{n=1}^{\infty} n^2 x^n = \frac{x}{(1-x)^2} + \frac{2x^2}{(1-x)^3} \quad (|x| < 1).$$

7.10. Expand the functions $\dfrac{1}{(1-x)(2+x)}$ and $\dfrac{\log(1+x)}{1+x}$ in power series.

7.11. Prove that if $f'(x) = f(x)$ for all $a < x < b$, then $f(x) = Ce^x$, where C is a constant. Deduce that
$$e^x = 1 + \frac{x}{1!} + \frac{x^2}{2!} + \ldots + \frac{x^n}{n!} + \ldots$$
for every x.

7.12. Prove that if $0 < a < 1$, then the equation
$$1 + \frac{x}{1!} + \frac{x^2}{2!} + \ldots + \frac{x^n}{n!} = ae^x$$
possesses in the interval $x \geqslant 0$ exactly one root $x = x(n)$. Show that $\lim_{n=\infty} x(n) = \infty$.

7.13. Prove that if $\lim_{n=\infty} a_n = g$ (finite), then
$$\lim_{x=\infty} e^{-x} \sum_{n=0}^{\infty} a_n \frac{x^n}{n!} = g.$$

Exercises on § 8

8.1. Let
$$f(x) = x^3 \sin \frac{1}{x} \quad \text{for} \quad x \neq 0, \quad f(0) = 0.$$
Investigate the existence of the second derivative $f''(x)$.

8.2. Prove that if a function f possesses the second derivative at a point x, then
$$f''(x) = \lim_{h=0} \frac{f(x+h) + f(x-h) - 2f(x)}{h^2}.$$

8.3. Prove that if the n-th derivative $f^{(n)}$ of a function f is bounded in a finite interval $a < x < b$, then the function f is bounded, too.

8.4. Prove that if $|x| < \frac{1}{2}$, then the approximate formula
$$\sqrt{1+x} \cong 1 + \tfrac{1}{2}x - \tfrac{1}{8}x^2$$
gives the value of $\sqrt{1+x}$ with an error less than $\frac{1}{2}|x|^3$.

8.5. Applying the Maclaurin formula to the function e^x evaluate the first four figures of the decimal expansion of the number e.

8.6. Expand the functions $\operatorname{ar\,sinh} x$ and $\sin^4 x$ in power series.

8.7. Deduce the formula
$$|x| = 1 - \tfrac{1}{2}(1-x^2) - \sum_{n=2}^{\infty} \frac{1}{2n} \cdot \frac{1 \cdot 3 \cdot 5 \ldots (2n-3)}{2 \cdot 4 \ldots (2n)} (1-x^2)^n$$
for $-1 \leqslant x \leqslant 1$.

8.8. Let f be a function continuous in an interval $a \leqslant x \leqslant b$. Prove that if its second derivative exists inside this interval and $f''(x) > 0$ everywhere, then
$$f(\alpha a + \beta b) < \alpha f(a) + \beta f(b)$$
for any positive numbers α and β satisfying the condition $\alpha + \beta = 1$. Likewise, if $f''(x) < 0$, then
$$f(\alpha a + \beta b) > \alpha f(a) + \beta f(b).$$
Give a geometrical interpretation of these inequalities.

Hint: Investigate the sign of the auxiliary function
$$g(x) = f(x) - f(a) - \frac{f(b) - f(a)}{b-a}(x-a).$$

8.9. Let $p > 0$, $q > 0$, $\dfrac{1}{p} + \dfrac{1}{q} = 1$. Prove that
$$xy \leqslant \frac{x^p}{p} + \frac{y^q}{q}$$
for $x \geqslant 0$, $y \geqslant 0$.

Hint: Substitute in the previous exercise: $f(x) = \log x$.

Exercises on § 9

9.1. Give an example of a function defined in an interval $a < x < b$, which do not possess a primitive function in this interval.

9.2. Prove the existence of a primitive function of the function
$$f(x) = \sin\frac{1}{x} \text{ for } x \neq 0, \quad f(0) = 0.$$

9.3. Find the integral
$$\int (|x-1| + |x+1|)\, dx.$$

9.4. Deduce the recurrence formulas
$$\int \frac{dx}{\sin^n x} = \frac{-\cos x}{(n-1)\sin^{n-1} x} + \frac{n-2}{n-1}\int \frac{dx}{\sin^{n-2} x}$$
and
$$\int \frac{dx}{\cos^n x} = \frac{\sin x}{(n-1)\cos^{n-1} x} + \frac{n-2}{n-1}\int \frac{dx}{\cos^{n-2} x}$$
(n is an integer > 2).

9.5. Find the integrals
$$\int \frac{dx}{\sin^3 x} \quad \text{and} \quad \int \frac{dx}{\cos^4 x}.$$

9.6. Applying the recurrence method calculate the integrals
$$\int \frac{dx}{x^n(1+x^2)} \quad \text{and} \quad \int \log^n x\, dx,$$
where n is a positive integer.

Exercises on § 10

10.1. Applying the formula
$$\int \frac{x^2}{1+x^4}\, dx = \frac{1}{4\sqrt{2}} \log \frac{x^2 - x\sqrt{2} + 1}{x^2 + x\sqrt{2} + 1} + 2 \arctan \frac{x\sqrt{2}}{1-x^2} + C$$
we obtain
$$\int_0^{\sqrt{2}} \frac{x^2}{1+x^4}\, dx = -\frac{1}{4\sqrt{2}} \log 5 - 2 \arctan 2 < 0.$$

This result is false, since the function under the sign of the integral is positive. Explain the reason for this error and find the actual value of the above integral.

10.2. Prove that if the function f is continuous and positive in the interval $a \leqslant x \leqslant b$, then denoting by M its upper bound the following formula holds:

$$\lim_{n=\infty} \sqrt[n]{\int_a^b [f(x)]^n dx} = M.$$

10.3. Let f be a function defined and continuous for all values x. Find the derivative of the function g defined by the formula

$$g(x) = \int_a^b f(x+t)\,dt.$$

10.4. Find the limit

$$\lim_{n=\infty} \frac{1^m + 3^m + \ldots + (2n-1)^m}{n^{m+1}}.$$

Hint: $\dfrac{2k-1}{2n} = \dfrac{1}{2}\left(\dfrac{k-1}{n} + \dfrac{k}{n}\right)$.

10.5. Evaluate

$$\lim_{n=\infty} \frac{1}{n}\sqrt[n]{(n+1)(n+2)\ldots(n+n)}.$$

10.6. Prove that if the function f possesses a continuous derivative in the interval $0 \leqslant x \leqslant 1$, then

$$\lim_{n=\infty} n\left[\frac{1}{n}\sum_{k=1}^n f\left(\frac{k}{n}\right) - \int_0^1 f(x)\,dx\right] = \frac{f(1)-f(0)}{2}.$$

Apply this formula to evaluate the limit

$$\lim_{n=\infty} n\left(\frac{1^m + 2^m + \ldots + n^m}{n^{m+1}} - \frac{1}{m+1}\right).$$

10.7. Evaluate the limits

$$\lim_{n=\infty} \int_0^\pi \sqrt[n]{x}\sin x\,dx, \quad \lim_{n=\infty}\int_0^{\pi/2} \frac{\sin^n x}{\sqrt{1+x}}\,dx.$$

10.8. Prove that if the function f has a continuous $(n+1)$-th derivative in the interval $a \leqslant x \leqslant b$, then

$$f(b) = f(a) + \sum_{k=1}^{n} \frac{f^{(k)}(a)}{k!}(b-a)^k + \frac{1}{n!}\int_0^b f^{(n+1)}(x)(b-x)^n dx.$$

(another form of the Taylor formula).
Hint: Proof by mathematical induction.

Exercises on § 11

11.1. Prove the convergence of the integrals

$$\int_0^\pi \frac{\cos\tfrac{1}{2}x}{\sqrt{x(\pi-x)}}\,dx, \quad \int_0^\infty \frac{\log x}{1+x^2}\,dx, \quad \int_0^\infty \frac{\cos\sqrt{x}}{x}\,dx.$$

11.2. Show that the integral $\int_0^\infty \frac{\sin x}{x}\,dx$ is non-absolutely convergent (and thus is an example of a conditionally convergent integral).
Hint: $|\sin x| \geqslant \sin^2 x$.

11.3. Prove that if the integral $\int_{-\infty}^{+\infty} f(x)\,dx$ is absolutely convergent, then

$$\lim_{n=\infty}\int_{-\infty}^{+\infty} f(x)\cos nx\,dx = 0 = \lim_{n=\infty}\int_{-\infty}^{+\infty} f(x)\sin nx\,dx$$

(cf. § 10.2, Example (8)).

11.4. Prove that if the integral $\int_a^\infty f(x)\,dx$ is convergent and the function g is monotone, bounded and possesses the first derivative continuous in the interval $x \geqslant a$, then the integral $\int_a^\infty f(x)g(x)\,dx$ is convergent, too.
Hint: Apply the Cauchy theorem (§ 11.1, 2), taking into account the second mean-value theorem (§ 10.7)

11.5. Prove that
$$\int_0^1 \frac{1}{x^x}\,dx = \sum_{n=1}^{\infty} \frac{1}{n^n}.$$

11.6. Prove that
$$\int_0^1 \frac{x^{a-1}}{1+x}\,dx = \sum_{n=0}^{\infty} \frac{(-1)^n}{a+n} \quad \text{for} \quad a > 0.$$

11.7. Prove that
$$\int_0^{\infty} \frac{xe^{-x}}{1-e^{-x}}\,dx = \sum_{n=1}^{\infty} \frac{1}{n^2}.$$

Hint: Apply the formula $\dfrac{z}{1-z} = \sum_{n=1}^{\infty} z^n$.

11.8. Let the function f be continuous and monotone in the interval $x \geq a$. Prove that the convergence of the integral $\int_a^{\infty} x^a f(x)\,dx$ implies $\lim\limits_{x=\infty} x^{a+1} f(x) = 0$.

Hint: Consider the integral $\int_x^{2x} t^a f(t)\,dt$.

11.9. Assuming that f is a function continuous and monotone in the interval $0 < x \leq 1$ and that the interval $\int_0^1 f(x)\,dx$ is convergent, prove the formula

$$\lim_{n=\infty} \frac{1}{n} \sum_{k=1}^{n} f\left(\frac{k}{n}\right) = \int_0^1 f(x)\,dx.$$

11.10. Let the function f be continuous, decreasing and positive in the interval $x \geq 0$. Prove that if the integral $\int_0^{\infty} f(x)\,dx$ is convergent, then the following formula holds:

$$\lim_{t=+0} t \sum_{n=1}^{\infty} f(tn) = \int_0^{\infty} f(x)\,dx.$$

In particular, find the limit

$$\lim_{t=+0} \sum_{n=1}^{\infty} \frac{t}{1+n^2t^2}.$$

Hint: Compare the proof of the Cauchy-Maclaurin theorem (§ 11.6).

11.11. Prove that if f is a function continuous in the interval $x \geqslant 0$ and $\lim\limits_{x=\infty} f(x) = g$ (finite), then

$$\lim_{t=+0} t \int_0^{\infty} e^{-tx} f(x)\, dx = g.$$

INDEX

Abel
    transformation 61
    theorem 60
Absolute convergence
    of an integral 309
    of series 68
Absolute value 19
d'Alembert criterion of convergence 63
Alternating series 59
Alternative 130
Anchor ring 271
Anharmonic series 59
Arc cos, arc sin, arc tan 88
Ar cosh, ar sinh, ar tanh 159
Area of a region 237
Argument of a function 81
Arithmetic mean 26, 49
Ascoli theorem 53
Asymptote 174
Asymtotic value of $n!$ 262

Bernoulli inequality 13
Bolzano–Weierstrass theorem 39
Bound (upper and lower)
    of a function 110
    of a set 20
Bounded
    function 97
    sequence 32
    series 58
    set 19
    variation 102

Cauchy
    criterion of convergence 63
    definition of continuity 104
    –Hadamard formula 137
    –Maclaurin theorem 294
    remainder 186
    theorems 44, 57, 71, 75, 155
Cavalieri formula 270
Centre
    of curvature 246
    of mass 246
Closed interval 82
Commutativity of a series 68
Composition of functions 94
Composed function 108
Conditional convergence of a series 68
Conjunction 130
Continuity
    Cauchy definition of 104
    Heine definition of 104
    left-side 103
    of a function 102, 103
    one-side 103
    piecewise 103, 281
    principle 18
    right-side 103
    uniform 109
Continuous function 102, 103
Convergence
    absolute of a series 68
    absolute of an integral 309
    absolute of a product 78
    d'Alembert criterion of 63
    Cauchy criterion of 63
    conditional of series 68
    logarithmic criterion of 298
    of integral 274, 277
    of a product 74
    of a sequence 29
    of a series 55
    radius of 123
    uniform, of a sequence of func-

tions 118
uniform, of a series 121
Criterion of convergence
  d'Alembert 63
  Cauchy 63
  logarithmic 298
  Raabe 67
Curvature 245
Cut 18
Cyclometric functions 88

Darboux
  integrals (upper and lower) 263, 264
  property 111
Decreasing
  function 83
  sequence 28
Dedekind continuity principle 18
Definite integral 224
Derivative 138
  generalized 179
  left-side 140
  logarithmic 158
  one-side 140
  right-side 140
  second 180
Differences quotient 138
Differentiable function 14
Differential 175
Direction cosine 244
Dirichlet
  function 92
  integral 287
Divergence to infinity 44
Divergent
  sequence 29
  series 55

e 47
Element
  of the area 240
  of the length 243
Elementary function 86, 161

Elliptic integral 274, 277
Entier function 83
Euler
  constant 180, 190
  formula 300
  gamma function 292
Even function 102
Existential quantifier 132
Exponential function 87
Extremum of a function 150

Factors of an infinite product 74
Finite increments (theorem) 153
Formula
  Cauchy–Hadamard 137
  Cavalieri 270
  Euler 300
  de l'Hospital 164
  Leibniz 173, 184
  Maclaurin 187
  Simpson 258
  Stirling 261
  Taylor 185
Fourier
  integral 226
  series 300
Fresnel integral 284
Function 81
  above and below bounded 97
  argument of a 81
  bound of a 110
  bounded 97
  composed 108
  continuous 102, 103
  cyclometric 88
  decreasing 83
  differentiable 141
  Dirichlet 92
  elementary 86, 161
  entier 83
  even 102
  exponential 87
  extremum of a 150
  Γ of Euler 292

hyperbolic 158
increasing 83
inverse 85
Lejeune–Dirichlet 92
limit of a 89
linear 86
maximum of a 149
minimum of a 150
monotone 83
odd 102
one-to-one 85
periodic 84
piecewise continuous 103, 281
piecewise monotone 83, 85
polygonal 127
primitive 201
propositional 131
rational 87
second degree of a 86
trigonometric 87
uniform continuity of a 109
value of a 82

Generalized derivative 179
Geometric
  interpretation of a function 82
  mean 26, 51
Graph of a function 82
Guldin theorem 251

Harmonic
  mean 26
  series 65
Heine definition of continuity 104
de l'Hospital formula 164
Hyperbolic function 158
Hypocycloid 272

Implication 131
Improper limit 45, 91
Increasing
  function 83
  sequence 28
  strictly 83

Indefinite integral 201
Indeterminate expression 163
Infinite
  product 74
  sequence 26
  series 55
Inflexion point 197, 198
Integral
  convergent 274, 277
  Darboux (upper and lower) 263, 264
  definite 224
  Dirichlet 287
  elliptic 274, 277
  Fourier 226
  Fresnel 284
  indefinite 201
  improper 273, 276
  Poisson 290
  Riemann 264
  seemingly improper 279
Integration
  by parts 206
  by substitution 206
Interpolation 256
Interval
  closed 82
  of convergence of a power series 125
  open 82
Inverse function 85
Improper integral 273, 276

Kummer theorem 66

Lagrange
  interpolation 256
  remainder 187
  theorem 153
Lateral surface 249
Left-side
  continuity 103
  derivative 140
  limit 91

Legendre polynomials 273
Leibniz formula 173, 184
Length of an arc 241
Limes 29
  inferior 53
  superior 53
Limit 28
  improper 45, 91
  left-side 91
  one-side 91
  right-side 91
Linear
  function 86
  independence 272
Lipschitz condition 117
Local property 105
Logarithm 87
Logarithmic
  criterion of convergence of series 298
  derivative 158

Maclaurin formula 187
Mathematical induction 12
Maximum of a function 149
Mean
  arithmetic 26, 49
  geometric 26, 51
  harmonic 26
Mean-value
  Lagrange theorem 153
  theorem in the integral calculus 229, 253
Method of indetermined coefficients 212
Minimum of a function 150
Monotone
  function 83
  sequence 28
de Morgan laws 130, 132

Natural logarithm 49, 87
Necessary condition 42
Negation 130

Newton binomal formula 15, 192
Normal
  region 239
  to a curve 141
Numerical line 11

One side
  continuity 103
  derivative 140
  limit 91
Open interval 82
Orthogonal system 273

Partial
  fraction 211
  product 74
  sum 56
Pascal triangle 54
Periodic function 84
Permutation 68
Piecewise
  continuous function 103, 281
  monotone function 83, 85
Point of inflexion 197, 198
Poisson integral 290
Polygonal function 127
Polynomials 87
Positive integers 11
Power series 123
Primitive function 201
Product of sentences 130
Propositional function 131

Quantifier 132

Raabe criterion 67
Radius
  of convergence of a power series 123
  of curvature 245
Rapidness of the increase of sequence 65
Rational
  function 87
  number 11

Real number 11
Recurrence definition 27
Remainder
   Cauchy 186
   Lagrange 187
   in the Taylor formula 186, 187
   of a series 56
Riemann
   integral 264
   theorem 71
Right-side
   continuity 103
   derivative 140
   limit 91

Schwarz inequality 17, 271
Secant 139
Sequence
   bounded 32
   convergent 29
   decreasing 28
   divergent 29
   increasing 28
   infinite 26
   monotone 28
   rapidness of the increase of a 65
   term of a 27
   uniformly convergent 118
Series
   alternating 59
   anharmonic 59
   bounded 58
   commutativity of a 69
   conditional convergence of a 68
   convergent 55
   divergent 55
   Fourier 300
   infinite 55
   power 123
   uniform convergence of a 121
Set

bound of a 20
bounded 19
Simpson formula 258
Solid of revolution 248
Stirling formula 261
Subsequence 38
Sufficient condition 42
Sum of sentences 130
Sum of the infinite series 55

Tangent to a curve 139
Taylor formula 185
Term of sequence 27
Theorem
   Abel 60, 125
   Ascoli 53
   Bolzano–Weierstrass 39
   Cauchy 44, 57, 71, 75, 155
   Cauchy–Maclaurin 294
   finite increments 153
   Guldin 251
   Kummer 66
   Lagrange 153
   mean–value (Lagrange) 153
   mean–value (in the integral calculus) 229, 253
   Riemann 71
   Rolle 151
   Weierstrass 39, 110

Uniform continuity of a function 109
Uniform convergence
   of a sequence of functions 118
   of series 121
Universal quatifier 132

Value of a function 82
Variable 82
Variation bounded 102
Volume 248

Wallis formula 259
Weierstrass theorem 110